The Applied
Scrum Master

The Applied
Scrum Master

Anthony Belau

ISBN: 979-8-9850959-0-6

For the next generations that deserve better

Contents

Chapter 1
Introduction

He is not the man you expected to see as he enters the room. The new construction of the building makes itself known with the self-assured click-thud of the heavy, well balanced door. The undetectable white noise throughout the office gives a dull feeling in the ears. The hushed chatter of the employees feels distant with the open concept office design. There are boxes on employee's desks from their recent move from their original office.

"Thank you for being here. I'm glad you could make it today." Talking to someone over the phone is different than meeting them in person. Knowing their name beforehand always constructs an image of who we expect to see. Michael stands to greet and shake hands with Chris, a director at FancyCat Technology for the company's sole software development team, Scrumbledor's Army.

"I hope the snow wasn't too much of a hassle last night. Let's get right to the point since my IT manager will be in shortly to speak with you. Our team is barely delivering half of what they say they can achieve each Sprint. Quality in the application is becoming a burden to everyone. It has become a death of a thousand cuts over the last few years. No matter what we do, nothing seems to change with them. Do you think you will be able to give this team some structure?"

"I certainly think so. My previous teams had similar metrics of about 50% completion of commitment when I started but we were able to average closer to 90%

towards our commitments after a few months of work. To be transparent, that figure isn't directly 40% more throughput of code delivered. There was a compromise of lowering commitments to be sustainable and then refining Stories into more granular pieces to achieve the improved consistency in completion and commitments."

"Okay, great. There is definitely a willingness in the business to compromise because the customers are getting pretty angry. What does success look like for you in your first week, if you were to accept a job offer?" Michael digests the question for a moment.

"I hope to open productive communication between the team members as soon as possible. If the team is failing to meet half of their commitments, team members are failing to communicate well somewhere along the process. It may sound like I'm oversimplifying the issue but open communication is fundamental for success."

"Good. Can you briefly give me an overview on Scrum and Agile? Just start from the beginning of what you know." Michael draws a breath as he immediately dives into his favorite topic that he has grown to love in the last few years.

About the book

The general communication about Scrum emphasizes the defined rules, instead of offering "why" or "how," especially for the Scrum Master role. This prompted me to reflect on my experiences to share insights I've learned along the way. This book is the same guidance I would provide in a mentoring session for what to expect in this role. I hope that you find this perspective valuable, or at the very least, challenging to what you already know.

There is no four year degree that prepares someone to become a proficient Scrum Master the same way there is for software developers. It also seems that no one knows about the role unless you're already working in Information Technology. Over the years as I've matured into the role, I've noticed an absence of applied knowledge of the Scrum Master role in the industry. From what I've seen, the vast majority of information in articles, blogs, agile coaches websites, or certificate courses, is predominately process oriented and not particularly helpful for how to be successful as a Scrum Master. These sources almost always suggest not to stray from the Scrum rules. My interpretation is that this is because sound Scrum knowledge communicates an authoritative and credible air with an assumption that to go against the Scrum dogma is a poor decision. However, you only need to work one day in the role to understand that things are not as simple as they seem.

The following is offered as an applied Scrum Master interpretation that has been meaningful and incredibly successful for me. My hope is that this may leave an impression on newer Scrum Masters looking for an applied perspective or people considering working for one of the many open positions where they can aspire in creating an empowering work environment for software engineers.

Please continue with an open mind and a healthy skepticism. Through your interactions with Scrum, or any Agile process, you will enter situations where the rules are unclear and best judgment must be used. The Scrum and Agile values and principles will serve as a general guide but, in my opinion, will fall short of the needs of a Scrum Master. It's like giving a chef the ingredients list but not the recipe for how to prepare it. I hope that the chapters in this book can be a source of confidence with judgment for various situations, as well as helping your team to become high performing and consistent.

I implore you to consider the concepts in this book. Even if you stop reading now, my parting words would be to keep a healthy dose of inquisitiveness by asking the questions regarding any process or meeting, "Why are we doing this? What is the purpose and value we should be receiving from this?"

Chapter 2
What is Scrum

No, this isn't about a Scrum in rugby, although the origins of the sport date over 2,000 years ago [1]. Scrum in the sport is considered a restart of a play. In project management, Scrum is a methodology to complete work consistently with short intervals of delivery. This chapter will summarize components of the methodology. Simply referred to as Scrum, it prescribes a set of values, meetings, and roles that constitute the structure. It falls under the project management theory of Agile as an iterative methodology that is driven forward by planning increments of time called Sprints. The result is a potentially shippable product by the conclusion of the Sprint.

The meetings are the Daily Standup, Sprint Planning, Sprint Review, and Sprint Retrospective. These four meetings can be collectively referred to as the Scrum Ceremonies. The three Scrum Roles, which make up the Scrum Team, are the Scrum Master (SM), Product Owner (PO), and the development team. The Scrum Values are commitment, courage, focus, openness, and respect. A Scrum Master will need to understand these key components.

Most Scrum practices in companies are leveraged to develop software. Since it is often the Information Technology (IT) department that depends on this process, the foundation of the references and expectations cater to this domain. Nearly every company or organization has at least a website with minimal functionality to provide informational articles, list of services, a "contact us" section, and a way of accepting payments. With this context, the result of a Sprint being a potentially shippable product is often assumed to be functional and tested software.

Each organization eventually has a tailored process that makes Scrum fit their needs. One size shoe does not fit all when applying Scrum practices, although when learning the methodology, it appears that there is only one right way to run it. Scrum is like an architect's guidance for drafting blueprints for a house. Every house has a roof, walls, flooring, windows, yet nearly every house is unique in shape, design, and add-ons necessary for those using it.

A Sprint is an agreed upon length of time not determined by Scrum. It is widely accepted to be two weeks and unchanging through a year but can range from one to four weeks. An argument against one week is that it ends up being a lot of Scrum meetings for each Sprint while more than two weeks can be difficult to consistently plan for and commit to at once. Work is planned, performed, and completed within the Sprint.

During the Sprint, the development team will have several User Stories committed to be completed. Team members can collaborate, work in pairs, or work independently in pursuit of completing a User Story. This contains a set of expectations called Acceptance Criteria. The recommended wording is the "Given, When, Then" format:

- Given (states the prerequisites of the scenario)
- When (identifies where it needs to occur)
- Then (lists the expected results)

- Given I have logged into the website
- And I have an active account
- And I have an active credit card
- When I type a letter or special character into the Credit Card Number text box
- Then an error message will display under the Credit Card Number text box
- And the message will have red text
- And the message will say "Invalid Credit Card Number. Please try again."

Velocity assists teams during Sprint Planning and guides the Product Owner when forecasting work in upcoming Sprints. Velocity is represented numerically as the average amount of work the team can complete in a Sprint. This is based on Story Points, which is an abstract estimate of effort on User Stories. There's more information on estimating and Story Points later on. Once the team has completed User Stories over several Sprints, the Scrum Master can look at the number of Story Points completed from the completed User Stories to get a sum for each Sprint. Then the average can be taken of the sums across several Sprints, which identifies a Velocity for that team. This is the relative standard to see how much the team can reasonably do insofar that the team remains consistent in commitment, composition, and similar work continues to head their way.

The Scrum Team Roles:
Product Owner
Scrum Master
Development Team

The Scrum Team consists of three roles. The Product Owner is responsible for what gets done by making decisions on what the results of work should be. They rank the backlog, define acceptance criteria, and are the representative of the business but not a manager of the team. The Development Team is responsible for getting the work done and helping the PO break down the User Stories into smaller, technical deliverables. It consists of cross-functional team members, however depending on the industry there may be some specialized roles within the team. In software development, a team will consist of software engineers, quality assurance, and potentially business analysts or graphic designers. The Scrum Master is responsible for how the process is applied while giving power over decisions to the Scrum Team. They are provided no authority and must enable the flow of work against the inevitable, sweeping range of impediments that will arise.

The Scrum Master is a servant-leader and coach. Servant-leadership is a leadership style that is expected to handle the needs of the team. The role as a coach is to provide guidance on Scrum, train the organization to leverage the methodology, and to coach the team to become independently functional where team members trust and rely on each other. I find these two expectations a careful balancing act. Would a football coach be on the field and directly throwing the ball for the quarterback? Would a basketball coach make free throws if his player gets injured? No, but they're also not servant-leaders. It'll be your decision as Scrum Master for when to be directly involved to resolve something and when to coach on an issue to independent resolution.

The best form of coaching is to identify and enable opportunities for the team to form through your indirect action [2]. The weakest form is direct involvement constantly without incurring team growth [2]. The Scrum Master is in the unique position to enable growth in those around them such that they can be better than they were the day before. It is often the case that anyone can wear that hat, however it is explicitly in the Scrum Master's job description to persistently pursue this. How you serve their needs and concerns, whether by becoming a middleman in communication, training on a task, or coaching on the Scrum values so they can decide for themselves, is entirely up to you as the designated servant-leader. I strongly encourage starting with observation and a light hand to see who steps up to meet the team's needs independently or what situations arise, then get directly involved to avoid complete failure or chaos. Unfortunately, you won't be their Scrum Master forever so the best path forward is prioritizing team formation and supporting individuals so that they can tackle whatever comes their way. These decisions are framed by the responsibility of applying the Scrum process as opening doors for the team to step through and improve themselves.

The Scrum Ceremonies:
Sprint Planning
Daily Standup
Sprint Review
Sprint Retrospective

Non-Scrum Ceremonies:
Sprint Demo
Refinement
Prefinement

Consistency requires emphasis at the beginning as it provides an initial foundation from which to measure and grow everything else. This is evident when estimating User Stories, determining Sprint commitments, adhering to adopted team practices, and tuning into sustainable effort in each Sprint. At Sprint Planning, which signifies the beginning of a Sprint, the Scrum Master will identify the team's velocity. The team can begin to determine how much they are able to commit to based on their velocity. They can then adjust their capacity based on team availability with time off, holidays, or anything that may impact the team's ability to complete committed work. Coming out of Sprint Planning, the team will have identified their capacity, considered any risks that may affect their successful delivery of work, and committed to as much work as they are comfortable with to confidently complete and deliver by the end of the Sprint. Consistency during Sprint Planning is where success begins.

The Daily Stand Up is the next Scrum Ceremony that occurs within a Sprint. This occurs every morning typically with a duration of 15 minutes, except on Day 1 of a Sprint, the day with Sprint Planning. The purpose of this meeting is for the team to comment on what was achieved yesterday or what they did, synchronize on work intended for today, and then discuss impediments. This is not a status report to management or to the team. It's the Scrum Master's responsibility to see that managers do not create this dynamic, intentional or otherwise. It's a matured expectation that team members will call out who or what they will need to continue the progress on their User Stories, or if their ability to contribute to the Sprint commitment is impacted through PTO, reassignment, or critical work that cannot wait.

Anthony Belau

Refinement is not a Scrum Ceremony however it's integral for productive communication between the team and Product Owner (PO). This meeting is owned by the PO but can be supported by the Scrum Master if the PO is new. The purpose of this meeting is for the PO to get their User Stories estimated by the team, which includes getting their input, questions, and feedback to understand the scope of expectations. When estimating with Story Points, the intent is not to simply put numbers on the User Story. It is to achieve shared knowledge about the User Story through conversation. To help achieve this, focus on reaching consensus within the team for the size. Any differences in sizes are prompts to drive conversation. Information will be shared where needed with this focal point. A quick revote with a majority win will keep the meeting moving forward when consensus is not an option. The theme from the Scrum Master should be "What questions do you have that may affect your estimate of this Story?" A knowledgeable PO will drive this on their own or eventually learn from your example.

The technique used for estimating Stories is critical for consistent results. Estimating, interchangeably called sizing, a User Story is the act of gauging effort to complete the User Story based on the written Acceptance Criteria. The standard technique for estimating is Planning Poker, which uses the abstract Fibonacci sequence. The numbers available for Story Points are 1, 2, 3, 5, 8, 13, 21, where 13 and 21 are typically too large as the unknown elements and risks of a Story grow exponentially where definition doesn't exist yet. The Planning Poker method is performed by showing the team a User Story's Acceptance Criteria, give them a minute to silently pick a size or ask a clarifying question, then everyone reveals their numbers on a count of three.

This will remove conformity when people are uncertain of what their estimate should be and that's exactly where transparency is needed to resolve gaps in knowledge and fight assumptions. By framing the conversation to focus on what work is necessary to move a Story from Ready to Done, it can be associated that effort is a relative number. This does not allow a conflation of time with effort needed. Everyone shows their estimate at the same time, conversation occurs about differences, and re-estimating occurs until consensus or majority wins.

Another method is relative sizing. Take two Stories that the team agrees are true to size and compare unestimated Stories to those by identifying if it's larger, smaller, or the same. Some teams use a single point of reference, say a 3 point Story, but this often causes the team to size only between 2, 3, 5 and rarely 1 or 8. This is easily avoided by using two reference points of 2 and 5, so the team is much more likely to use the whole range of the Fibonacci Sequence available to them.

Sprint Review will be more difficult to get value at the cost of the team's time in attendance in this meeting. Ideally this meeting has a PO present that sees the demonstrable features that have been completed and provides immediate feedback, positive or negative or even new requirements. This meeting can also involve a review of Sprint metrics with the team, Scrum Master, Product Owner, and Stakeholders. This is an opportunity for the team and Stakeholder to connect on what is being delivered for their estimated Story Points. Bear in mind that metrics are always just the start of the conversation, nothing more, nothing less. They provide information but not enough to make effective decisions by themselves. It's good practice to only show any metrics as a percent, instead of a numeric value. It can make identifying a successful Sprint more clear to Stakeholders, who are likely to be unfamiliar with the process and jargon. Not all teams benefit from Stakeholder engagement, not all teams will need to do a Sprint Review. It has been my experience that this meeting is the one that strays the most from Scrum ideals.

The Sprint Demo's purpose is to present the completed, functional product to the Product Owner. There is the added benefit of walking through all the changes with the team, since not everyone will always be on the same page for what is finally being delivered by their coworkers. Not all teams have to do a Sprint Demo. If the Product Owner is engaged enough through the Sprint, it may be a waste of time to go through the structured meeting unless a Stakeholder, or the business, is engaged and interested enough, which I rarely see. Although it may not be a valuable meeting by itself, as a Scrum Master, it would be in the team's best interest to at least try the meeting once or twice to be familiar with it and the technology available to the team, should the meeting be needed.

The Sprint Retrospective has the most value potential. If you don't do Scrum, or are doing any process ineffectively, consider adopting some form of this meeting. The team comes together to discuss what went well, what didn't go well, and what they can do to improve. This will identify actions from the team to start relieving pain points or getting movement towards goals that they have always had but never initiated. Without effective Retrospectives, the team is subject to iterative stagnation. Team members should feel empowered to voice their concerns regarding anything and they should be reminded that the process is in their hands to modify as they deem best.

Prefinement is the meeting you have never heard of but have always wanted. This is not a Scrum Ceremony. This meeting came around as a result of one of my previous team's needs to bring the team closer to the PO earlier in the project's inception while decomposing requirements. It's a solution and design meeting, like a Refinement without the sizing. The meeting's intent is for the PO to bring in a few of the team members to review Epics, which is like the project's idea, and assist in drafting Stories as necessary. Although it's not an expected output of the meeting, this also makes some team members available to discuss potential sizes, if they're familiar and comfortable with their current Story Point sizing process of the team.

There are three Scrum Artifacts. These are the locations of information about the product or projects. The Product Backlog, which is referred to just as the backlog, is a single location owned by the PO where User Stories are ranked by the PO and sized by the team in preparation for a future Sprint. The Sprint Backlog is the location owned by the team that contains their current work for the Sprint. The Product Increment is the collection of potentially shippable products that is expected as a result of the successful delivery of Sprint. This Product Increment should not need to go to another Scrum Team to have work continued on it so that it may become functional.

With the core concepts of the overall process laid out on the table, it's important to note that the existence of these meetings and roles doesn't mean that Scrum is being done well. The prescribed meetings are a set of tools that should help leverage a set of values and empowerment for teams. If the meetings are there but not an understanding of the values, then the overhead exists

Anthony Belau

without the benefit. These are the tools available to build the house, it's up to you to work to define the best house to build for your needs.

"Work is always easier when someone else has to do it." - Mr. Noel.

Story

Fear creeps into his mind on the heels of lingering doubt that something is wrong. He tries to reorient himself on the streets of the strange new city. The curving roads and chaotic intersections make the map difficult to follow. Getting lost is not in his agenda but the anxiety is always there when starting a new job and wanting to be precisely on time. He thought he knew what cold was until he experienced his first winter in Boston, and the locals are calling this winter 'mild.'

Although feeling late, he arrived before the rest of his new team. The office is small and dated. The company had started in the same office 20 years ago. It shows in the drop ceiling tiles that have water spots from leaks, entire departments that outgrew their spaces, and a building that only blows hot air through the vents. He's glad to see that his desk is situated in the middle of the single IT team for the company. He hasn't met anyone from the team yet but soon he will see how much work is cut out for him as their first Scrum Master.

Michael's experience with a handful of small companies as a Scrum Master built his understanding of the process and, more importantly, the intent. Unfortunately, he has mostly seen a lot of the bad sides of Scrum due to previously attempted-and-failed transformations. This has led him to even see the ugly side of the process when things are misunderstood. This new team is nearly a clean slate. He is the company's first Scrum Master where he can set a strong precedence. It's why he accepted the position. It's a meaningful opportunity.

The door slams open and a true Bostonian walks in. Energy beams from the man like heat from the sun.

"What's up dude! You must be Michael. I'm Simon. It's wicked awesome to have you here." Michael stands and shakes his hand.

"I appreciate the opportunity and I look forward to working with the team. When is Standup?"

"9:30," Simon checks his smartphone. "In an hour."

"Okay, thanks, I'll introduce myself to the team after the meeting but I'd really like to observe everything first before taking over any facilitation, if that's okay."

"Sounds good. Get settled, you should have some papers on getting logged into your machine. Let me know if you need anything."

"Will do."

The hour was enough to get set up to become familiar with the login process and access new hire emails. There's enough time to complete a few of the initial training courses as the team trickles in and catches up over the weekend. They seem to all chat together except Anushka, who works remotely in India. 9:30AM EST is 8:00PM IST. Collaboration will take a more conscious effort to include the international member as a larger part of the team.

Simon dials into the meeting on the polycom. They say their good mornings and good evening to Anushka, who works later in her local time to be available for the company and is nearly finishing her day to have an overlap with the US people. Simon launches into the meeting.

"We finally have our new Scrum Master that I've been talking about all last week. Thanks again Michael for joining us. He said he would like to observe our process first before taking over any meeting facilitation. That said, let's get rockin'. I spent a bit of time on..." The standup continued for twenty five minutes with the time spent going into technical discussions and monologues that did not concern the team but only concerned the member or team lead. Another issue was the roll call aspect where Simon was calling on each team member individually to speak, as if he were a teacher of a class. This was not a standup of a fully enabled team that takes ownership of their own meetings.

First impressions are the most important, particularly for a Scrum Master providing a tone of expectations. "Thanks for giving me the opportunity to work with y'all. I really hope we can make some positive changes and find where we want to grow as a team, and commit to making those improvements. Please be aware that as a Scrum Master, I am not a manager nor an authority, I'm simply here to help us become more effective at meeting our commitments and improving. If you have any questions, please don't hesitate to ask. This will be more conversational than training oriented. I will not be mandating anything for our process as much as possible. I will try to guide you to the answers I'm hoping for and provide suggestions with justification where necessary."

The QA engineer raises his hand.

"What are you going to do that's different from what Simon already does?"

"That's a great question." This question, and those to follow, are very significant as they will tell Michael exactly how much understanding each team member has of Scrum and its application. "I am not technical. Simon will still be your technical SME and manager. I am responsible for our Scrum practices. I will be taking over Scrum meeting facilitation, providing guidance on the process, holding us accountable for continuous improvements, and removing any impediments to your progress or growth. As a 'coach' I will be interacting with you as a team and individuals to see that we're communicating effectively. Scrum has a lot of rules but it's important that we understand the intent behind said rules."

The same hand is still up in the air. "What if we don't want to do Scrum? I've heard a lot of bad things about it and I happen to like our current process."

"Well, if you truly don't want to do Scrum, I hope that we can revisit the topic one-on-one so that I can better understand your concerns for creating an environment where you will thrive. Scrum can have a negative reputation and it is not perfect. I want to emphasize that I am not here to drive towards doing Scrum by the book. I am here to ensure we're delivering software that brings value with a consistent cadence. I hope that we can leverage some of the valuable processes that Scrum provides towards this objective. If we try something, and it truly doesn't work, then we will disband the process. Does that sound fair?" A quick nod said that he was satisfied with the answer.

Simon thanks Michael again and the meeting ends. Everyone goes right into their routine for work but Simon approaches Michael. "You handled that well with Erik! Everyone is excited but a little nervous. They don't know what to expect from you so I'm glad you spelled it out. Just a heads up that we have our Retrospective in a little bit since it's the last day of our Sprint. Don't be late!"

Michael reflects on how the standup went. His mind walks through each person's personality and communication style. He wonders who would be open enough to speak their mind to him and who might need a little more time before being honest with feedback to the Scrum Master. Were all the questions appropriate during standup? Why not? Why are the technical discussions going on for so long? Does this annoy the rest of the team or is it just going to be part of their culture? These are the necessary questions that Michael will keep in mind as he continues to observe the team. Fortunately, his favorite Scrum meeting is right around the corner.

The Retrospective in the afternoon begins, however Anushka is not able to dial in since it's late evening locally. It does not seem like she sent any notes for the team to discuss from her perspective. This will be something for Michael to address sooner than later to begin including her more into the team. He begins noting as much as possible on what is, and isn't, being communicated during the Retrospective. Simon begins this meeting as well.

"Okay, let's do this. Let's cover what went well, what we can do to improve, and next steps." No one takes notes. After a few moments of silence, a junior engineer named Liu speaks up.

"Well I think our code reviews are improving."

"Yeah, we're actually doing them this Sprint," chimed in Jose, another junior engineer. The team snickers but it didn't feel like genuine laughter. The sarcastic tone says that code quality is an issue but they know it shouldn't be. Michael wonders how many production issues are in the queue for this team to handle. Liu continues a few moments later.

"We finally finished up the data migration for the new line of business. That dragged on forever." The team laughed again with more genuine relief. Simon shifts the focus.

"Yeah, yeah, we don't have much that went well. What can we do to improve?" The team sits in silence. Michael knows that this is guiding the team away from a productive Retrospective.

"May I make a suggestion?"

"Of course."

"Instead of asking for improvement ideas, can we further discuss what didn't go well, even if we think we can't fix the problem?"

"Sounds good. Team, what didn't go well?"

Erik's hand went up immediately. "I have something. I keep getting Stories late in the Sprint. There's not enough time to manually test and we've no automated testing built to help. I have no idea what we can do to improve this but I'm having to stay late in the afternoons to get my work done. It's pissing me off."

Simon replies, "What if we had some sort of internal stage gate in the Sprint? Anything that isn't done coding by Day 7 gets pulled out which would give QA time to focus as the Sprint nears its end." Michael knew that the team should probably discuss the root cause a bit more but they've a solution that is worth pursuing so he holds his opinion. The team nodded in agreement with Simon's idea. The meeting continued with much of the same tone until Michael's most skeptical team member raised his hand again.

"Hey Scrum Master, Scrum seems to be relatively clear on its rules. What do we do when Scrum doesn't have a definition on what to do?" Michael thought for a minute then spoke.

Anthony Belau

Chapter 3
What is Agile

The principles of Agile are demonstrated within several methods of development, such as Scrum, Kanban, and Scrumban. Agile is the approach of streamlining the flow of work from concept to delivery on an iterative timeline. This contends with "Waterfall" as the primary theme in software development organizations. Waterfall is considered a linear development process and not iterative. Waterfall is all or nothing for project progression. It is considered rigid, time ineffective, and difficult to adjust priorities within. Agile provides a newer theory when approaching work and sets the measure of progress as working software in an iterative delivery cadence. This enables companies to create a work environment that keeps up with the ever-evolving needs of customers and clients.

The Agile Manifesto is the heart of it. This defines the separation from traditional delivery practices. It's four statements where the concepts on the left take priority over the statements on the right but the concepts on the right still represent value. These don't apply perfectly in every scenario but they are a sturdy gauge for making decisions along the journey when things become unclear for determining the better path towards an Agile environment. These statements are not expected to be applied in a black and white fashion. You will find yourself standing between the two when determining the appropriate degree of action frequently as a Scrum Master.

The Agile Manifesto

-Individuals and Interactions over processes and tools.
-Working software over comprehensive documentation.
-Collaboration over contract negotiation.
-Responding to change over following a plan.

"**Individuals and interactions** over processes and tools" is the first statement. This stands most starkly against older development environments. This is the vision for communication with everyone in an Agile environment. The Scrum Master will need to be a champion for sparking individual interactions, encouraging it whenever possible. Working with **individuals** means accountability. **Interactions** encourage strong understanding in a two way conversation where questions get answered, or at least immediately presented to the other person, and expectations communicated. Processes and tools are undeniable in their necessity for an efficient organization but to rely on them for most communication is setting people up for failure. It's too easy to see different interpretations based on text or within a structured process where the constraints are defined by the tool but the needs of the person are vastly more complex than the tool will likely allow explanation within.

For the Scrum Master to navigate this statement effectively, it should be encouraged to close communication gaps immediately where possible and practical. The lines of communication within teams, with their dependencies, and within the larger organization should be considered for potential gaps. I suggest an immediate phone call if someone has a question about their task at hand. If that's not possible, then a chat message that leads to a call is still preferential to sending

Anthony Belau

an email. If it's important, I also encourage follow up chat to an email, even if the email is required due to the complexity and structure required for the information. Why schedule a meeting for next week if a fifteen minute call can solve it now? This begins to shape the team's immediate surroundings to be more responsive to their needs, allowing them to react more quickly to questions and risks as they should arise more quickly, and hopefully, begin to make them stronger impediment resolvers.

This also means if a concern needs escalation, do it sooner than later. Do not wait on sending an email or updating a ticket somewhere then hoping for the best. If something is needed, try to bring it up before it is actually needed so that people can start helping earlier. Bringing people in to assist instead of relying on a process to resolve something will create an environment that aligns more with the Agile values, particularly this statement. Of course there are exceptions to the rule for when processes should be strictly followed instead of reaching out to an individual but it should be recognized that this will not make a company more Agile with this kind of dependency.

This goes beyond communication as well. An individual's aptitude matters more than the tools used. Bug-finding software can be helpful as an aid but it should not be expected to replace technical skill, time spent refactoring, nor the adaptation of additional best practices or improved code review guidelines. Aides like these should define minimum tolerances, not that maximum effort to achieve acceptable performance. Individuals and their opinions need to be recognized over the tools themselves. There are arguments that validation tools provide peace of mind to clients, which is fine, but that is not the maximum that should be expected, it's the minimum.

"**Working software** over comprehensive documentation" is written for organizations that have software as a component of their product or as the product itself. This can be generalized as "functional product over comprehensive documentation." This breaks away from Waterfall where all requirements and designs are expected to be agreed upon with the customer before writing code begins. This can traditionally take months or the better part of a year to solidify based on what I've heard. Documentation becomes overhead since it needs to be maintained after any changes are requested. Companies that want to be more flexible and competitive with the market, will need to work without complete documentation in pursuit of competitive agility. **Working software** makes money, comprehensive documentation does not, in nearly every instance.

"**Collaboration** over contract negotiation" is the signal to see if communication has turned sour between the development team and Product Owner. By being aware of the differences in tone, the Scrum Master is able to help people understand how this can affect communication and motivation. When there is low trust between the team and Product Owner, discussions over User Stories can easily slip into contract negotiations where the Product Owner pushes their known Acceptance Criteria without adjusting based on the team's feedback, causing the team to simply accept it and not apply critical thinking to what is being presented nor its technical faults. When the tone shifts into **collaboration**, where the Product Owner is actively eliciting feedback, the team will be engaged in working towards the best product when their concerns and thoughts are at least considered. This shift in intent can be cultivated within the team itself rather than a tone shift in the Product Owner first.

Software development is a creative endeavor. Collaboration is a highly productive way of working to reach the best answers. Collaboration increases motivation, results, and overall happiness with better experiences for knowledge workers. Contract negotiation has an implication of meeting minimum expectations to the letter of what is written. This is stressful since written communication is frequently open to interpretation and may not make sense for the individual that is expected to perform that task and then become ultimately responsible for any decisions they may make. If the culture is to do exactly what is written, then the results will be less valuable, innovation will be nonexistent, and motivation will be at a minimum of just doing enough to get by every day.

"Responding to change over following a plan" provides guidance for how the organization and individuals should handle requirement changes. Following a plan is comfortable since the direction is known and the expectations are already understood. Responding to change means restarting the collaboration process on a brand new idea. Applying technical expertise requires diligent and concerted effort to be effective. This becomes taxing if too many changes are coming in. The only reason anyone has a position in organizations is because it ultimately provides a benefit to the customer. The project providing the most benefit to the customer sometimes changes, sometimes very frequently. Sometimes within days or weeks. The needs of customers have only evolved more rapidly and it shows no signs of slowing down. Being flexible to respond to change will often bring the most value over following a plan.

These four statements are the scale for weighing decisions and actions to determine how balanced a path forward is when in an Agile environment. The Scrum Master should be aware of how much weight to give both sides of the manifesto's values and be prepared to facilitate discussions about which actions may be best for the team. The more frequently they're referenced, the more familiar with applying them the team will become. Some Scrum teams may never mention the Agile Manifesto due to a lack of any reinforcement. This is the foundation of expectations for any of the Agile methodologies. Some things are unable to be influenced and the most that can be done is to remind the team that we may fall more heavily on one side than is ultimately best.

Do not enforce the Agile Manifesto completely nor belittle decisions that lean on the right side of the statements. Teams and organizations are principally driven by doing what's necessary in an attempt to avoid any unnecessary work, which can include going against a new methodology's expectations. It takes years to establish an Agile environment. This takes time because adaptation to a methodology has a learning curve and can involve slowing down delivery before benefiting from the agility. A little understanding from the Scrum Master can go a long way when helping teams create a positive environment as they change over their ways of thinking and working.

Supplemental to the Agile Manifesto, there are the 12 Agile Principles that help further define the vision of Agile. Most of them can be taken at face value without additional context. There are some I think benefit from elaboration. These are numbers 5, 9 and 11 as they provide more substance for the Scrum Master in communicating with team members. They are great to introduce to new engineers joining a team or when you're joining a team.

12 Agile Principles

1. Our highest priority is to satisfy the customer through early and continuous delivery of valuable software.	2. Welcome changing requirements, even late in development. Agile processes harness change for the customer's competitive advantage.	3. Deliver working software frequently, from a couple of weeks to a couple of months, with a preference to the shorter timescale.
4. Business people and developers must work together daily throughout the project.	5. Build projects around motivated individuals. Give them the environment and support they need, and trust them to get the job done.	6. The most efficient and effective method of conveying information to and within a development team is face-to-face conversation.
7. Working software is the primary measure of progress.	8. Agile processes promote sustainable development. The sponsors, developers, and users should be able to maintain a constant pace indefinitely.	9. Continuous attention to technical excellence and good design enhances agility.
10. Simplicity–the art of maximizing the amount of work not done–is essential.	11. The best architectures, requirements, and designs emerge from self-organizing teams.	12. At regular intervals, the team reflects on how to become more effective, then tunes and adjusts its behavior accordingly.

My favorite to highlight first is number 5, "Build Projects around motivated individuals. Give them the environment and support they need, and trust them to get the job done." This might sound like common sense, yet companies struggle to effectively support and give power to teams. Building projects around motivated individuals is demonstrated when a problem exists and someone offers a possible solution. They then should be enabled to see if they're correct by investigating and bringing a more complete solution to the table. Once they've had time with their solution, they can help break down the components of that solution and help others understand the vision as well. This is ownership without delegation, which is motivating as people will voluntarily step up to the plate when responsibility is available and they feel comfortable with the task.

For example, a car is broken in a repair shop and you're a Scrum Master facilitating the car's repair. Three engineers are standing by speculating as to why it's broken and what needs to be done to fix it. Each person may have a different starting point for troubleshooting. If you were facilitating repair, the best way you could foster motivation would be to enable them to pursue their own initiatives first after a brief collaboration and coordination. How might motivation change if you were to tell them what to do first instead? In this simplified example, the engineers are already thinking about their individual troubleshooting starting point. They just need to be unhindered in applying themselves. The argument against this example would be that in most cases, direction is needed for effective use of time, and that's absolutely true. However this is just an example to show how enablement might foster motivation. Put yourself in their shoes. Could you see that your outlook might change if you were the engineer in the example?

Trust is fundamental to having motivated teams and driven engineers. No one should be micromanaging in an Agile environment, unless there's a critical issue at hand that requires immediate coordination. Scrum Masters need to be trusting and encouraging as often as possible. Coach for self-reliance, not compliance. A team manager, while not officially considered part of the team, will have incredibly valuable input at Refinements. They often have concerns with the meeting if it occurs without them, however they should be able to trust the team to get the work done right. Sprint Planning should continue if the Product Owner is unable to attend as well, since we should trust that the Backlog is ranked and reflects what the Product Owner wants with detailed Acceptance Criteria. This takes cultivation from the Scrum Master to achieve understanding on both sides and self-sufficiency for the team to work without external support, regardless of the value the support can bring. This task of fostering trust will never stop as new team members join.

Next is number 9, "Continuous attention to technical excellence and good design enhances agility." The pursuit of technical excellence is the result of effective teamwork and individual contributions. Good design is yielded from the same actions. I have yet to see an engineer draft a completely perfect design for a complex application on the first pass, without critiques, feedback, or review even with decades of software development experience. When the team culture is collaboration focused and the expectation is technical excellence, the team will indeed have enhanced agility because they will be able to leverage each other's strengths while building in incremental design that allows improvements to the feature in the future. This allows for simpler and sounder logic to be written, tested, then built upon later with increased confidence and reliability.

Technical excellence is almost always a goal for companies, regardless of the methodology or environment. Driving discussion based on this ambition helps the Scrum Master bring conversations into being open about opinions and motivation within the team. An engineer should be able to speak their mind if they have a concern, doubt, or see a flaw. The pursuit of this ideal hones the team's ability to deliver the best possible value. It means that we should evolve our team's culture and those interacting with them in pursuit of excellence. The engineer is the steward of the application. They should not simply work on the code as if it belongs to someone else. They should feel ownership of the code they write and carry a sense of pride that comes with their work done well.

Last is number 11, "The best architectures, requirements, and designs emerge from self-organizing teams." Of the three emphasized, this is the most difficult to bolster since most of the self-organizing opportunities are often withheld from the team. This will take constant advocacy to not let this be forgotten by managers and stakeholders. By encouraging the team to identify adjustments towards self-organization where they see fit, and the Scrum Master working to identify the opportunities as well, this can bring concepts into discussion that were once assumed as the norm and thought unchangeable.

Initiative is distinct from motivation. It's integral to the will to be self-organizing and is an absolute must have from engineers. This idea reinforces principles number 5 and 9. My perspective is that trust must exist before initiative and technical excellence can grow. Without trust, initiative and technical excellence become dampened and discouraged. The goals of technical excellence and self organization through initiative are critical to get the most motivation within a development team. This all becomes unlocked as potential only once trust is established.

There is a theme in IT where team leads or managers will take on too much responsibility by demonstrating too much initiative by unintentionally taking these opportunities from the team. The very attributes that got them into the higher level position, are the same traits that will hinder opportunities for other engineers to step up and to have more responsibility. By highlighting this relationship on both sides, you can begin to draw better boundaries for the team to grow within and push against.

Numbers 1, 2, 3, 4, 8 and 12 are self-evident.

1 - "Our highest priority is to satisfy the customer through early and continuous delivery of valuable software." The customer is broader than the people who are paying. Customer in this statement includes anyone consuming the product, whether internal or external to the organization. Early and continuous delivery of valuable software identifies the methodologies that fall under Agile.

2 - "Welcome changing requirements, even late in development. Agile processes harness change for the customer's competitive advantage." Changing requirements can be painful to deal with for the development team. The optimal path is to work towards whatever is most valuable and if changes reflect that, then the team should try to adapt. If changing requirements are constantly received by the team late or

mid-Sprint, then there should be discussion to understand why this is happening afterwards. It should be evaluated to mitigate impact on the Sprint's commitment. Otherwise this impacts the team negatively and wastes time with work that is in progress during the Sprint. The issue is felt by the team, raised by the Scrum Master, and often resolved by the Product Owner that actively protects their backlog from unnecessary changes from upper management instead of capitulating to every request.

3 - "Deliver working software frequently, from a couple of weeks to a couple of months, with a preference to the shorter timescale." This has become an industry standard. Companies are severely behind the curve if they're still releasing quarterly. The value is being able to receive customer feedback on a product earlier to adjust accordingly rather than committing to work towards an idea received poorly by customers.

4 - "Business people and developers must work together daily throughout the project." This manifests itself when a Product Owner is not overloaded and is available to attend functional demos of User Stories in the middle of the Sprint to give immediate feedback or approval. Sometimes requested work comes from someone other than the Product Owner. The developer should have the freedom to demo to whomever will be using the software as it's created.

8 - "The most efficient and effective method of conveying information to and within a development team is face-to-face conversation." This is the question and answer cycle being greatly reduced and cannot be reduced any further via another form of communication. People are able to immediately reiterate their understanding to drive comprehension. Barring a detailed

email on specific topics, conversations are absolutely key to conveying information. As companies shift into remote work environments, the second best form of communication is video calls with the camera on. People can tell when someone is listening or not quite understanding or if they're distracted on their phones. Roughly 80% of communication is body language and nonverbal.

12 - "At regular intervals, the team reflects on how to become more effective, then tunes and adjusts its behavior accordingly." This is embodied by the Sprint Retrospective. Teams need to be benefiting from this, not just going through the motions. If this meeting is performed to simply check a box that it was completed, the team will not grow to its full potential. Issues will reoccur, frustrations will reoccur, gaps in communication will continue. The team needs to benefit from their Retrospective and that responsibility falls squarely on the Scrum Master.

The last three remaining, numbers 6, 7, and 10, warrant some context.

6 - "Agile processes promote sustainable development. The sponsors, developers, and users should be able to maintain a constant pace indefinitely." This needing to be an explicitly identified principle is indicative of the nature of companies. Development teams are composed of knowledge workers, not factory workers. They focus on the applied use of knowledge, not just the use of rote labor. This is a critical distinction because it's easier to mindlessly move your hands in a repetitive task than to think through stress, mental exhaustion, and brain fog. This has a direct impact on the work-life balance too. It's a common experience to start

slurring words, forget words and thoughts, and be ready to fall asleep hours ahead of a regular bedtime due to busy days. This principle's purpose is achieving predictability and sustainability.

The pursuit for sustainable pace will be a constant challenge. Companies often lose sight of the amount of effort and raw time that goes into delivering high quality software. Most will want the software delivered quickly instead of with quality. The highest quality often being an afterthought although they will not admit it as they boast an adoption of Agile to clients. The pressure for speedy delivery can eventually become overtime or working on the weekends which can become a norm for the team. The Scrum Master should constantly be aware of this pressure and keep the team working at a stable pace. It has been my experience that new teams will be very agreeable and will unlikely resist outside pressures independently until they have become more mature.

7 - "Working software is the primary measure of progress." This comments on carrying incomplete work into upcoming Sprints. If a User Story is not completed by the end of the Sprint, the entire User Story and points should be carried forward into the next Sprint. Though you may be asking, "Why not just split the Story to accurately reflect how much was done in each Sprint?" This will be tempting as it makes the team feel reassured that all of their effort is accounted for, as well as making it easier for the Scrum Master to show consistency within metrics. This is poor practice because it undermines the need for proper refinement and commitments. By enabling carry over and Story splitting, it removes the pressure and the need to be applied as much. A half developed User Story that is still untested is not working software. This also signals to the team that metrics are

more important than they should be. It's acceptable to have a dip in points in a Sprint as long as it can be explained and spark corrective action. If we didn't deliver functionality by the end of the Sprint, the team needs to reconsider how they're sizing, their practices, and practicality of the commitment from Sprint Planning.

10 - "Simplicity - the art of maximizing the amount of work not done - is essential." Simplicity being essential is straightforward. Every action on a daily basis should be kept as simple as possible. This means reducing the frequency of switching between tasks. For example, in Refinement, it's simpler to just update the acceptance criteria in the middle of the meeting during discussion instead of having the Product Owner update the acceptance criteria after the meeting. It's better to spend 15 minutes discussing a User Story before development begins than to spend a day readjusting code later. It's better to understand if a current feature is expected to be more configurable and reusable later. This way can be written right the first time and not reworked later, possibly falling on another engineer or team that has no knowledge of its original design.

The Agile Manifesto and the 12 Principles form the tenets of an Agile environment to create teams that are highly adaptable. The Scrum Master will need to be well versed in these concepts, the intent, and their application. Development teams, Product Owner's, and stakeholders sometimes understand the concepts intimately and even so, they may still miss the intent of the message or have used poor practices for too long. Not all of these principles may be achieved completely every Sprint but it's the vision and pursuit of these goals that grows a team into a better version of themselves every Sprint.

Story

Michael responds to Erik's question about what to do when situations are not neatly accounted for in Scrum's defined rules by elaborating on the Agile Manifesto and Principles, what it means to be Agile, and leveraging the core values to provide the basis for judgment when things become unclear. The team appreciated the honest answer that the process is not perfect but there is no clear instruction for the team's application of Scrum and Agile.

The day ends quietly as the team's Sprint is over. They will begin Sprint Planning the next day. Michael's heart sinks when he sees the meeting invite is from 10:00am-3:00pm, with an hour lunch in the middle. It's a four hour meeting. This doesn't bode well for how Refinements are going, although he hasn't seen one yet. He sighs as he shuts his laptop to conclude his first day. Team members will not be motivated and prepared for a Sprint after a grueling four hour meeting.

The gusts of wind between the 150 year old buildings bite the cold deep into his woefully unsuitable jacket for Boston's mid-winter weather. It had worked so well in Florida's winter. Walking while making fists in his pockets in an effort to keep them warm, he imagines the history of the city. Horses and carriages roaming the same streets and a port bustling with Irish accents. It still amazes him to learn that Boston tripled the available footprint through landfilling throughout its history. He walks along Summer street to South Station to take the commuter rail home.

The next day arrives and Michael is early again. This gives him time to begin walking through what he might expect from the lengthy meeting and to prepare for any curveball questions that might come his way. With a team this size, he doesn't expect Sprint Planning to require more than an hour at worst. If the backlog is prepared and the Product Owner is present, thirty minutes sounds more reasonable. Michael sets this as his ideal for where he would like to see the team.

The office door slams open. Simon, the ray of sunshine, announces that it is a good morning as he sips his Dunkin' iced coffee. Michael takes the opportunity to get some information from Simon before anyone is around to hear, hoping that this encourages Simon to speak his mind freely.

"Hey Simon, why do you think this team needs four hours to plan a Sprint?"

Simon shakes his head disappointedly. Clearly this isn't his vision for the team either and possibly sees that there's room for improvement.

"We go into each User Story, make sure everyone has a detailed understanding, then we go into breaking it down into Tasks and estimating the hours. I know it's time consuming but that's what the meeting is intended to do. I think the team also lacks confidence about the Stories, so by going through it all together, everyone feels more prepared. Since everyone is content with it. Well, almost everyone, we haven't been able to improve it yet."

Michael nods in agreement, taking all the details in. They are missing some of the core values as engineers that will enable them to reach higher aptitude and agility. There's no ownership on the individual if they break everything down into tasks in hours as a group. Michael fears that the team's skill level will become stagnant if they are not comfortable taking any risks.

"Got it. I saw Standup on the calendar for this morning. Normally with Scrum, we do not have Standup on the day we do Planning. Is it okay if we cancel it or do you think the team needs the meeting this morning?"

"Yeah we can cancel it. I'll let them know, thanks." Michael gives a short nod and a smile and then sits down to continue becoming familiar with what's needed from him on an administrative level. The team trickles in and Simon lets them know about the cancelled meeting. No one seems to be disappointed by the small change.

Sprint Planning comes around and there's a stillness that falls over the room. No one chats before since there's plenty of talking to be done soon. The tension of going into the meeting could be cut with a butter knife and served on rye toast. Simon leads the meeting by sharing on a wall-mounted TV so everyone can see the Backlog and Sprint Backlog. The rest of the meeting is a slog. Conversations have long technical tangents, the team doesn't reach consensus on estimates and some simply give up during debates, the Product Owner is missing due to too many meetings, and Erik is pushing back constantly because he has concerns of how much he will be able to manually test, even with the new Day 7 gate the team intends to try.

During a brief break at the 3 hour mark after lunch, Erik approaches Michael and asks the question that everyone is thinking.

"Hey Scrum Master, this meeting is brutal. How do you think you will be able to help us deal with all these rules in place when you've no authority? We seem to always add more but never take any away, like taxes." All sidebar conversations fade as all eyes turn to Michael. He stands up, takes a breath, then he provides his thoughts. He expected a curveball but didn't expect to get this into Scrum on his second day.

Chapter 4

Values over Rules

Rules establish predictable results when making a decision. Values allow a variable result when making a decision. Values over rules sounds like the default in companies but often it's still rules over values, regardless of what's signaled from upper managers. A mindset of values first is the only way to create lasting change as a Scrum Master. Practicing rules over values permits bad practices to be unchallenged, inefficiencies to persist, and lowers accountability across the board. Rules are easy to change while values are not. It is difficult to work under micromanagement after you've worked independently and responsibly. The precedence under which teams should operate will need to be demonstrated by the Scrum Master. Scrum states that all Scrum teams will demonstrate courage, openness, respect, commitment, and focus. Demonstrate these values on a daily basis if there's any hope to leave a lasting impression.

Organizations convey a set of values that are expected of their employees which defines the work culture that is supported by leadership. Scrum Masters should be no different when trying to establish an Agile culture. As a servant-leader and Agile champion, it has been my experience that the team will emulate actions, speech, and words as you set the new team norm for openness, courage, and focus. It is important that team members are not scorned for inconsistencies in behavior nor told to mirror the Scrum Master's actions. Simply embody the values and be consistent. Try to find opportunities for them to grow into the behaviors. This is not an overnight change.

Courage is facing a challenging situation. There is no shortage of demanding tasks for software development teams, considering legacy system bugs, undocumented processes, and an ever increasing demand for features balanced against sound quality. As a Scrum Master this may look like dealing with a manager you strongly disagree with on approaches or an uncompromising stakeholder that wants the team to do overtime to meet a low priority goal. Being able to say "No" and drive a healthy conversation takes a tremendous amount of professional confidence and personal courage from a new employee. By demonstrating this value, you are paving a path that will encourage the team members to speak up as well. Courage for engineers can manifest itself in speculation about designs or admitting ignorance of a topic to receive help. It's agreeing to new challenges and learning outside of a comfort zone, which often stagnant teams will resist by continuing to employ old practices.

Courage is also in the willingness to take risks. The environment will need to be built where team members feel comfortable taking calculated risk. This isn't to be reckless. This is to allow innovation and to push against what may be assumed to be the most efficient or effective methods. For example, teams can be like race cars. They're often pushed to their limits with the pedal to the metal until something breaks, losing the race. Teams typically have pit stops that are not a large enough part of the development process. The pit stops cost valuable time in the race and the results are not a new, shiny feature or car. However, if a team were to take a risk and are allowed a predictable pit stop to refresh, improve, and upgrade before any break occurs, they'll go faster after or at least not break, causing a predictable loss in the race. Risks that team members avoid tend to things like needing to reassess what is required for a Story, being incorrect on a technical aspect, or being perceived as a lesser engineer for a perceived failure. Work with leaders to remove this pressure and encourage team members to take these risks. Without this, the team will never become high performing.

Openness is voicing opinions and concerns. This is essential for collaboration in pursuit of a high-performing team and technical excellence. The driver for openness in discussion is to ultimately determine what is successful for the team. Scrum works when everyone leaves ego at the door. The only ego, camaraderie really, should be the pride in being in an exceptionally Agile team that can handle anything that comes their way. This works when the environment is also one that allows for risk taking. If the Scrum Master is unable to communicate openly, the team will never communicate openly. The only goal is value and a high quality product. Any success or failure that leads us to that goal is desirable.

Respect is not judging someone for having a differing opinion or perspective. This is necessary for enabling teamwork and growth. Respectful culture produces a risk-free zone to perceptions and judgments about an engineer's aptitude. This atmosphere is highly motivating and causes less turnover when engineers feel understood by their teams. Feeling like a valuable contributor is substantially different from feeling like an outsider that can't speak candidly.

Commitment is seeing agreements through to completion. This value is the north star for consistency across Sprints and spurs the team to estimate User Stories with reliability and accuracy. This foundation builds trust within the team and with the business. Successful commitments sparks motivation as it's the fruit of hard labor. As the team begins to grow by meeting commitments, it shows that the team was able to accomplish something they would not have been able to do previously. If the Scrum Master is unable to change anything else at the organization, this should still be the guiding value in hoping to align with Agile.

Focus is getting the task at hand moving forward without wasting time. This is a component of Scrum because there are plenty of small tasks that need to occur quickly, efficiently, and routinely so that the process may experience more flow and throughput. Software development requires extended periods of focus to be able to foster understanding and to productively move a Story forward. As a Scrum Master this is demonstrated by giving the team your undivided attention, which will invite the team to mirror you with other team members. This sounds like common sense, and it is, but there will be plenty of temptation to multitask while working in IT. Sometimes if someone is not focused, you are in the position to be able to call people out, but only if needed. Be mindful that team members will sometimes be handling critical items during the meeting, so that can easily backfire if you're heavy-handed in your attempted correction.

Rules can become questionable when examined through the lens of the value that should be received. Rules could be that a specific person always has certain duties like design work, resolving certain problems, approvals, or communication. How would communication flow and our environment change if we were to be more open about concerns and provide sincere feedback? If we were to offer respect and the benefit of the doubt in the face of mixed performance, how might we engage that employee and support them? If we aimed for commitment every step of the way, how might our motivation change or communication improve? How might meetings be more productive if we spoke up with courage to volunteer information rather than sitting in silence?

By exhibiting the Scrum values, you are signaling to the group how they should behave in an Agile environment. As they see you speak your mind, say "No" when needed, and apply the other values, they will see that it is okay to speak up as well. It's important to set your expectations with the process with the team and then be consistent in your demonstration. By using the right phrases, the team will begin to use the right phrases as well. By framing your arguments respectfully and productively, the team will begin to do so as well. I suggest never emphasizing this behavior or asking the team to mirror you. Simply conduct yourself in an Agile manner and they should perceive then adapt if they agree or see the benefit.

Rules are the backbone of any organization. I am not suggesting that these rules are unnecessary nor that they all be broken down. Alternatively, be wary of the ones chosen to challenge. A given organization has, hopefully, been successful up to this current point. To question the established rules is to question the foundation on which their success is built. However, what has made a company successful in the past are likely not the same principles that will allow the company to continue to experience growth by continuing to meet market demands.

Expecting to change existing principles is no small feat. Rules are not written on a wall waiting to be revealed. Rules for a team can be unchallenged assumptions about how things occur and how information flows within the business. My intent by emphasizing values over rules is that we should consider the validity of a rule, official or unofficial. We need to consider if it hinders the team's growth in becoming self-organizing, cross-functional, and adaptable. Is there an opportunity to trust and add responsibility to our engineers rather than rely on a manager's approval?

Example Rules:
- Only Team Leads perform code reviews.
- We do not change acceptance criteria on Stories ever.
- We should never reach out to another engineer directly.
- Your Agile Coach wants you to change your team's meeting schedule.
- When the boss makes a decision, it's final.
- Standup only begins when the PO finally joins.
- Refinement does not begin without the Team Lead
- Consensus in estimating is necessary every time.
- Only the person with the most experience will design workflows.
- No one is allowed to challenge the senior engineers' design.

Can you think of some you have experienced?

Anthony Belau

Identifying and taking note of these rules is a constant task. This becomes second nature as the team's interactions are observed. The goal is not to break all of these rules or have an agenda against them. We want to be cautious and considerate when weighing rules against values. If we have too much initiative and no rules governing individuals promoting code freely, there will be merge issues, overwritten work, and potentially bugs in production that will affect clients and ultimately revenue to the company. This will land people in hot water before you can say "It worked when I tested it locally though."

Servant-Leadership

An Agile transformation will have conflicts between the Agile values and the established rules set within the organization. Trust is impaired by micromanagement and poor communication. Technical excellence is impaired by lack of opportunity and lack of collaboration. Initiative is impaired by too many meetings on the calendar and, again, micromanagement. Scrum is offered as a solution to other management practices by empowering a team of capable individuals. The person that protects this empowerment is a Scrum Master that employs Servant-Leadership.

Servant-Leadership is everywhere when you learn about Scrum, read a blog about it, or go for a certification. This term is able to be defined in different variations for different applications. As it relates to Scrum and the Scrum Master, I define it with the driving theme as "Are my actions best for the team?" You are not strictly a servant to the team that will do every single task asked without question. Servant-leadership can be directly involving yourself to remedy a situation or training someone to solve their own impediment. Both approaches are acceptable and expected as long as the action can be judged to be best for the team.

The job description says that the Scrum Master will motivate the team. This does not mean that you're yelling at people to deliver their User Stories faster or that they need to pick up the slack so that they can exceed their Velocity by 5% this Sprint. This means that you're highlighting successes and bringing a positive attitude every day. Some teams will have a culture where recognition and congratulations are the norm. Some simply will not greet each other. I think that to motivate team members is akin to fully enable them to pursue their ambitions. Engineers have typically spent at least four years studying their field and are always continuously learning about technology. The motivation for creative problem solving is intrinsic for most engineers. As a Scrum Master that has influence through the organization and trust of managers, you need to enable them to fully pursue what they think is best. What's more motivating than autonomy and responsibility? What gets someone out of bed in the morning more than purpose? Imagine working in an environment where you had a hand in shaping rather than just a cog in a giant organization?

We can always correct code and handle late work in upcoming Sprints, however we cannot take words back. It's necessary as a Servant-Leader to ensure you're projecting the guidance and vision that the team needs and expects of you, as well as maintaining the direction that you need the team to grow in. This is why I suggest choosing Values over Rules.

Story

Cool mist and rain drizzles from above. Florida has intense downpours for fifteen to thirty minutes. Boston is different. The light, day-long drizzle of cold rain had replaced the biting cold. At least his neoprene Northface jacket is serving him well in the spring season here in the new city.

The team is making strong strides on growth but there are still some issues impeding their ability to continue to become a high performing team. He hasn't discussed these points with them yet but he has had conversations with Simon to ensure that their vision aligned. Consistency in routine is great to ensure that the team is meeting their commitments and goals, since growth and change have a way of causing delays. Unfortunately, this private equity business is not truly interested in high quality. Their goal, although unspoken, is the most value with the least possible effort. This has caused Simon to become a bit overloaded.

"It's not that I don't trust them. I just don't think they're ready."

"To be fair, you're not helping them grow by not giving them an opportunity to step up and take ownership of the system."

"I've seen their code. It's not pretty." Simon laughs. "I might get fired if anyone actually took a look at what's holding this system together!"

"Of course it's not quality, they expect you to catch all their errors since you won't let anyone else do code reviews. Why should they bother to put in effort?"

"I see your point. We've improved so much though but something still doesn't feel right. Do you think we're doing Scrum yet?"

"We're definitely on our way. The team is using the jargon accurately and asking more in-depth Scrum questions about scenarios they're running into, which is great. I think we've run our course on solving some of the immediate issues the team was facing but we need to start thinking on how we're going to look in the long term."

"What do you mean?"

"Well, it's good we're doing code reviews. It's bad that you have the sole responsibility for it. I don't think the team will bring it up. Would you mind bringing it up in the next Retrospective this afternoon? It'll be more impactful if it comes from you. It would be interesting to hear what they say."

"I'll bring it up but I'm not excited about the outcome. I hope you're right about this."

"Just imagine what we could become."

Chapter 5
Growth over Routine

"An organization that treats its programmers as morons will soon have programmers that are willing and able to act like morons only." - Bjarne Stroustrup

The hallmark of Scrum is growth in lasting improvements, which results in increased value through the team. If this is not occurring then you are not iteratively improving as much as you could with this methodology. It's applying the overhead but not realizing the benefits. Routine is the backbone for a team's sustained success. Yet it should not be prioritized such that actions sustained are good enough. We should constantly be reassessing what we're doing, why we are doing it, and what we can do to improve or at least stop doing to make it worse.

Retrospectives provide the primary channel for garnering ideas for potential growth. This meeting doesn't need to focus only on what didn't go well and then deciding what to change. What went well and seeing how things can be done better next time is also a viable path in the meeting. This is a good problem to have, however the Scrum Master should always work to maintain a broader perspective for everyone.

For example, I worked with a team that had assumed a best practice for code reviews before I joined. The team lead for a team of seven engineers, printed the sections of new code in hard copy, wrote out critiques in pen, then sat with them individually to walk through the changes needed. This is definitely an enriching practice to train and grow engineers but it's not sustainable for the team lead and becomes a hard cap on throughput with this process while also dissolving potential for other engineers to critique each other. In our retrospectives, we discussed how this process impacted the team. I withheld my opinion so that the team may define their own experiences and as a result, will be more driven to see it through.

The team eventually arrived at peer code reviews, as expected. Although risky to change, it was necessary to attempt growth. I assured them that we could always go back to the original process we knew if the new one fails to be effective. The team adapted the peer reviews and never looked back. Rather, they continued to build on and refine the peer code review practices by defining more standards amongst themselves, communicating when it's needed, and being available to conduct them. Nearly all of these discussions occurred during the Retrospective meeting where as the facilitator, I would guide these conversations through productive steps forward and support their decisions from manager over-ruling because of the risk.

If a team is in a comfortable routine and has not changed at all for years, they are stagnant and are becoming outdated and less effective with each passing day. Similar to economic inflation impacting a personal savings account, where inflation is the larger IT industry and the savings account is the knowledge and processes of the IT team. Every year, the IT industry is growing and expanding in all directions, so an IT team needs to grow to even keep pace with the industry. If a team is unchanging in this changing market, they are actively falling behind. This is the subtle decay of organizations that haven't adapted over years, while ignoring that the investment for updates to catch up becomes much steeper. There are more organizations than you'd expect that are several generations behind in technology. Some still rely on paper release documentation where free automation tools are the modern goal.

The advantage of iterative small changes instead of large sweeping changes in a given window is well understood in business, health, and education; respectively as PDCA [3], weight loss [4], and studying a language. Plan, Do, Check, Assess. The risk is lower when any sort of change occurs in small increments. Large changes have broader scope of uncertainty. I recommend that the Scrum Master direct as few changes as necessary, coach on the Scrum and Agile intent so that team members can adjust on their own, and based on any action items coming out of the Retrospective.

The existence of growth is encouraging since it indicates an unfulfilled potential. Part of identifying where potential exists is by listening to the concerns of the team members or management. Concerns are rooted in some sort of dissonance between ideal expectations and reality. Listen to what is being spoken, or even indicated by body language, in these complaints and consider possible paths forward. Initial concerns often need more understanding of the issue, however this will typically be on your own initiative to pursue. Other concerns may be rooted in unattainable expectations, therefore trying to find a compromise is as good as it gets for a remedy. Some concerns may be because of the context or a one-off problem where you're not needed. No one is going to narrate or deduce these out for you on their own accord. It's on the Scrum Master to take the time to understand what is impacting the team and to come up with plans to ameliorate them.

A team I had newly joined made an off-hand comment of "I just wish we could get into this new Azure DevOps stuff for our release process." I challenged this stagnation by following up to understand how long they've been waiting to start the transition. It had been a year without any sort of traction or intentional effort. I asked why this was the case and the team unwittingly knew the answers. It was because they didn't know where to begin nor set the time aside for it. They assumed that it would take time outside of the Sprint commitment to make progress on this infrastructure item that doesn't directly generate revenue with a customer, although the impacts are profound to the stability of the platform and relieving stress on engineers.

In Retrospectives, I asked the team their thoughts on negotiating with the business to set aside some capacity and discuss how Scrum might be leveraged to get the traction needed. They agreed so they discussed the need of this with the newly minted PO to see if we could set aside some capacity for team prioritized work. She negotiated with the business stakeholders and we compromised with 10% of our velocity to work towards whatever we deemed a priority. With this opening, the team decided to create Enablers, from which they would create User Stories and get true traction on improving infrastructure. By the time I left the team 18 months later, they went from a manual build and hand merging files to a nearly complete automated deployment process. More importantly, the team learned how they can negotiate with the business and use Scrum to make the improvements they need with sustainable expectations.

This is an experience and perspective that can positively impact the engineers for the rest of their careers.

Fostering ownership is another component for team growth. If individuals are not contributing into the workflow process, it will fail, regardless if it's Scrum or another methodology. Having individuals and team members accept Scrum as the mechanism for change and decide to leverage it to their benefit will ensure that growth begins and sustains without any regression. Cultivating acceptance is solely dependent on the Scrum Master's capability to demonstrate, coach, and facilitate to reveal the benefits. Once this begins to occur, the Scrum Master will have a basis from which to show that progress can be made and to encourage the team to continue and take more ownership of the process themselves.

Consistency and patience are necessary to support an upwards trajectory. Results will not occur within a Sprint. Changes should not be forced by the Scrum Master. They'll need to be discussed and then enabled in the right circumstance, allowing team members to make decisions that align with the goals at a time that's best for them. Set expectations for 3 two-week Sprints at a minimum before beginning to look for any measure of growth. Set your expectations to 13 Sprints, half a year, to begin to see any permanent changes within the team. It is likely without constant reinforcement and buy-in from the team on the new changes, they will revert back to their previous habits when the process comes under stress or if the Scrum Master role is removed.

Skills of a Scrum Master

In addition to the demonstrable values for Scrum, there are soft skills that are appropriate for a Scrum Master to grow and utilize to be more successful. The skills are communication, empathy, insight, initiative, organization and integrity. Through application of these skills as a servant-leader, the values of Scrum are more readily seen, understood, and mirrored by teams. Confidence in coaching ability provides weight when your recommendations are offered. Having a clear vision of these skills every day will ensure you are sending a consistent message and model for the team to emulate.

What does communication as a skill look like though? It's active listening techniques, facilitation techniques, and prudence in meetings and interactions. When someone is a savvy communicator, they are able to more easily understand what another person is trying to say. Even if they're doing a poor job explaining, the skill involves asking clarifying questions, paraphrasing, and sometimes offering open ended questions to help understanding for all involved. Be aware of who is in the audience, whether you're presenting or not. Consider their perspective on topics and potential reactions. Ask clarifying questions if you don't understand, or even if you know the answer but suspect someone else may not. For every question asked, several others are wondering the same thing. Questions can be a very powerful tool for the Scrum Master [5].

6 Active Listening Skills[6]:
Pay attention
Withhold judgment
Reflect
Clarify
Summarize
Share

Facilitation Skills:
Be prepared
Be flexible
Manage time

Don't be afraid to ask for clarification. "I'm seeing this acronym everywhere. What does it mean?" This one looks silly when written but there is no shortage of acronyms and no shortage of assumptions that everyone understands what they mean. Information is rarely spoon fed in Information Technology. Otherwise try restating the content back to the speaker. This is a great way to build understanding. "May I try to repeat back what I'm understanding so that you can correct anything I'm misunderstanding?" Also consider your audience and their perspective. "Is anyone not familiar with how this application functions at a high level? I'm about to get into the details of one specific part." Provide opportunities for two-way communication. "Does anyone have any questions?"

When communicating, consider that some teams are international and may not understand some slang or metaphors used. For example, telling a US engineer to "not wear a red shirt on his first day" has a good chance at being understood as a silly reference to a Star Trek antic where the security guards in red shirts are always the first to die in the show, thus dangerous to show up on your first day in a team with a red shirt. If you say this same reference to a woman from India, she may think you're specifically telling her what to wear. Be aware of your audience and be prepared to inclusively explain a joke if necessary.

Empathy is the ability to understand and sympathize with others. This is necessary as a facilitator and coach when you're trying to view things from another perspective. It helps you predict questions, concerns, and criticisms, as well as understand the mood of the room, team, or meeting. This skill is natural to most but it can be learned as well. Most importantly a little understanding goes a long way [7]. Try taking the Myer-Brigs personality test yourself and consider what other people's personalities may be to gain some insight into how others might think. Try viewing the other person's perspective based on their education, role, previous work experiences, and culture. What might they expect from those around them? How do they view you? Why? This will help you prepare to handle scenarios as they arise.

- How would you try to get an introvert to talk more?
- How would you try to calm down a situation that is escalating, considering their positions in the organization?
- Would you know if the motivation is down and how would you build it up?
- Would you know when it's an appropriate time to add some jokes?
- Is now actually a good time to provide a Scrum explanation or is the team working on something critical?

Insight is the deduction of what transpired in a meeting or a conversation based on observations and current knowledge to create a deeper understanding. This skill is used when sifting through perceptions to determine the implications. This can be as mundane as conducting a training session on providing basic Scrum where the team has zero questions. I would leave the meeting aware that zero questions were asked. This could mean little interest in the content, lack of comfort in asking questions either because of my being new to the team, current team dynamics, or company culture. I would actively try to provoke more questions going forward. Only as a last resort would I consider doing the most terrifying thing a facilitator could do to an innocent team member. That is to call on someone in the meeting to repeat what they heard so you can confirm their understanding.

Take the time to reflect during the work day. You are not expected to be hammering out spreadsheets or focused on intensive tasks all day, every day. Take the time to think through situations and come up with possible responses. Based on my observations and reading the room, I'll provide feedback to managers if I feel morale is low for a specific reason, or if someone has been assuming more responsibility lately. This takes active comparison of the past with the present to understand changes or shifts, particularly if they're subtle.

Initiative is required for almost any job but it is core for a Scrum Master's ability to inspire others. This role is typically unsupervised and is provided leeway for strict accountability, partly due to the difficulty in measuring results and impact. For a role that coaches how Scrum is to have work planned and measured, it's comical that a vast majority of the work is dependent on the Scrum Master's ability to initiate, progress, and complete tasks and projects without structure, supervisor follow up, guidance, or support. It's important to keep yourself accountable for your time that's spent outside of Scrum meetings. In the likely absence of guidance, define your own vision for near future work. Based on what you've seen or heard, particularly from the Retrospective meetings, what could you do to benefit the team? The most important result from your efforts will be simply demonstrating initiative to provoke it in the team. Enthusiasm and motivation can be contagious, so become a font of these traits. Don't be afraid to tell your boss or team if you're becoming idle. They will appreciate the openness from you. They may also become more open with you, possibly revealing some interesting ideas to be investigated or areas you could be of assistance.

The willingness to use the skill of organization is an absolute must have for the Scrum Master. It's often the case that the servant-leader will be handling a mess that others don't want to deal with, whether it's organizing documents or figuring out what is actually occurring with a given situation. Positions in IT are often in the midst of constant chaos for everyone involved, whether you're a manager, developer, QA, or in the business side. The Scrum Master will likely step into a role where there is a need to organize their own work, reliably establish accurate metric collection, and organize several things for teams as well. Assisting the team could look like documentation repository curation, ticket queue delegation, work management tool optimization, or helping channel communications coming in. It's likely that the team has charged ahead amidst critically shifting priorities and they've left organizing and documentation for last, if any documents exist at all.

Time management falls under organization. Practice and experience develops a keen sense of the pacing for meetings and conversations to determine when interjecting is appropriate. Keep a clear focus in your head so that you are able to keep track of technically complex conversations. For meetings that are exceptionally technical and outside of your aptitude, it can be like a maze that you are navigating blind folded while maintaining your bearings through the maze as the questions in the meeting turn left, right, and backwards without a clear sense of substantial progress. By maintaining bearings through these types of meetings, you can still provide time management to keep things on track.

"The commander in the field is always right and the rear echelon is wrong." - Colin Powell

I think this quote highlights the integrity expected of a Scrum Master, as well as providing confidence in applying judgment without guidance. In my opinion, this role is not quite a typical employee in the traditional sense. The Scrum Master is not often provided explicit tasks and then graded on how well or quickly they can be achieved. The role is a servant-leader and a coach. This has similarities to being a junior military officer. They both carry an enormous amount of responsibility and trust from those above but won't always get their hands dirty with the work the same way everyone else on their team. This confers an expectation to apply independent judgment for the direction of the work. Sometimes in hindsight, there will be wrong decisions and wrong things said at times. However, as long as things are moving forward the best they possibly can with the information at hand, you're functioning as an excellent Scrum Master.

The primary skills expected of a Scrum Master are communication, empathy, insight, initiative, organization and integrity. Those working in Scrum are expected to demonstrate courage, openness, respect, commitment, and focus. As rules set expectations, these are often built on what is known but these need to be challenged and improved for the betterment of the application for the team, organization, and customer. By focusing on the values, we can create and begin to work towards a vision of the ideal, not just sustaining what our organization says we should be doing. Therefore, we should maintain Growth over Routine as the Scrum Master.

Story

As months pass, the cold warms to a chill and snow becomes rain as spring prepares to bloom from winter. The dreariness of the seasons weighs on the team, however the longer days and sunlight definitely help to keep spirits up. It helps that every morning the Christmas lights are still up in the IT room's ceiling for all of Boston to see.

The team continues to grow and improve upon the action items that are identified during Retrospectives. They've also been successful in sustaining growth in practices that are already going well. Liu is exploring improving their manual build process by leveraging devops. They have continued to expand their code review process by doing peer code reviews.

Everyone has left for the Friday afternoon except for Simon and the Scrum Master. Michael approaches Simon.

"I'm really happy with the progress we're making."

"Same here dude!"

"But I think we can do even better. I've been here for several months now and it seems that the devs are growing in responsibilities as they relate to the team needs, however it may be time to expand their capabilities to help with the wider company."

"What do you mean?"

"I'd like to see the engineers engaging the PO more during Refinements, and maybe they can take the lead on solutioning Epics or providing you designs for sign-off, without the initial hand holding."

"I see."

"This would free up your time more as well and open doors of skillset growth for both you and the engineers. Thoughts?"

"We can talk about it again. How about at our next Retrospective?" Michael nods in approval.

"Sounds like a plan!"

The meeting came around and the intent was well received. The engineers saw the reasons for allowing more growth. They are highly motivated individuals and technical learning is a constant in the industry, of course other opportunities for career advancement make sense.

The following week, one of the engineers is working on an Enabler Story and is nearing his first design ever. He cannot finish without some input from the business based on some unexpected results he found during his analysis. He needs direction on how to handle the situation.

"Hey Scrum Master, can you help me with this impediment?" Michael drops what he is doing and nods at Liu.

"What's the problem?"

"I can't finish Story without getting some questions answered."

"What questions?" Liu shows Michael his current list and who he'd need to answer them. Michael nods and says, "Okay."

Liu is confused. "Okay, what? Can you help me with this?"

"Of course."

"So can I email you this list so you can help me with this impediment?"

"Let's see if PO is busy and just give him a call. You can search for him just by typing his name. Ah, he shows available so just give him a call. He won't mind at all. He's probably expecting it."

"Okay, so I'll just call him."

"Yep! Let me know if you have any issues reaching him."

Liu is able to get all his questions answered over a twenty minute chat. The engineer reflects for a minute on how Scrum Master resolved the impediment with him.

Chapter 6

Coaching over Direct Involvement

Coaching and direct involvement are opposing approaches that will be balanced on a daily basis. This balance is expressed well in the phrase "Give a man a fish, feed him for a day. Teach a man to fish, feed him for life." Directly getting involved to resolve an issue will be highly valuable at times for the short term, yet the ability to coach and train others is even more powerful for the long term. The highest form of coaching is for a Scrum Master to not fulfill every teaching opportunity but instead having peers teach each other the skills necessary to resolve impediments. This will build an exceptionally powerful team that can overcome any obstacle on their own.

The Scrum Master for a single team should expect to be hands on with completing Stories or assist in any other way possible. This could be data analysis, following up on impediments on someone's behalf, debugging, manual testing code from the team, writing any documentation the team needs, sitting in on long meetings to then give the team info while having everyone else to skip it, and completing any administrative tasks the team needs done. The higher your aptitude and ambition to assist, the more the team will benefit but the results still rely solely on you and your ability to work autonomously. The team will rarely have time to invest in training a Scrum Master. So long as it's one team and one team only, this will be a fair expectation and great opportunity to become more technical. Expect to be more directly involved than to coach in this situation. Otherwise, expect to mostly coach when working with multiple teams, ideally two.

Regardless of the number of teams working with a Scrum Master, it's important for a coach to be aware that all teams will go through Tuckman's stages of group development [8]. Forming, Storming, Norming, and Performing. The last stage identified as additional is Adjourning. I think that only high-performing teams that have a strong identity will naturally want to Adjourn but for other teams, it's only a job to them. A team will cycle through the main four stages as any changes to team composition occur. The coach will need to be aware of how their interactions could change while navigating the stages. This will be a constant reassessment.

Forming is the initial stage. As the name suggests, this is where first impressions of team members amongst each other will develop. Teams also fall back into Forming whenever there's changes to their roster. People tend to be on their best behavior and won't openly criticize another person's idea. It will be necessary to provide guidance as well as strongly communicating initial Scrum expectations and Agile values. This stage lasts as long as everyone is unfamiliar with one another. Your role is to help break the ice and prompt wider conversation, if it's not thawing organically already. There are plenty of resources online to help facilitate this, or simply begin small talk by providing a detail about yourself, then asking others about themselves to relate with.

Storming is when the team begins to challenge internal and external boundaries set in place around the team. The Scrum Master helps by instigating the open challenging of ideas and working through compromises all around. This will likely include a chat with managers to empower the teams to have the freedom to identify changes they would like to see and to apply them without criticism. Without this spark of challenge, there can be no innovation and the team will become risk averse and complacent.

Norming is when the team begins to understand each member's strengths and weaknesses. They will be able to leverage that knowledge to support each other. The team will begin to communicate more openly in this stage and may even hold their own social events. The Scrum Master will need to listen to the chatter and be aware of what is occurring, even if it's not in a facilitated meeting. Be aware of miscommunication between teammates during a code review to be able to help if it begins to escalate. I wait until after the 3rd souring exchange before considering stepping in, only waiting to see if they will resolve the conflict between themselves. Step in if it becomes too personal or it sounds like the conversation may have a more permanent impact on their working relationship. The team needs to learn how to work together with each other, the best results being organic. A Scrum Master that intervenes too often or early will heavily stunt the necessary interpersonal relationship development that occurs through conflict within the team.

Performing is where everyone wants to be and where everyone thinks they are. The team will readily commit to, meet goals predictably, and is capable of accommodating significant last minute redirection. The team will have an identity and camaraderie, which further increases their pride in reputation and performance. It's where Scrum really shines with the Retrospective meeting, although the Retrospective is critical to bring teams to this point to begin with. We will want to ensure we are relentlessly growing in some direction. If a team is constantly unable to accommodate change but are performing well by company expectations, you are still expected to find ways to improve, which will likely involve trying to dial back the team's commitment. Teams in this trap will run well for a while, but once a team is accustomed to stagnation, it can be difficult to get the momentum of adaptation and learning again.

Adjourning is the closure that teams should want as the team is disbanding or restructuring drastically. One last social event should be held with everyone so they can catch up and exchange contact info and plans of where they might want to head next. Strong teams can forge relationships that last for many years to come. Properly dissolving the team will bring needed closure and forward thinking to each team member.

As a Scrum Master, you will help teams and team members individually navigate these stages as the team grows and changes over time. There is always comfort in doing what you know well but we need to be steadfast in constantly pushing boundaries.

In pursuit of creating a stronger team dynamic, providing a place for them to talk amongst themselves is crucial to establishing interpersonal growth opportunities and identity. Similar to the idea of "locker room talk," team members should be more free with their opinions, thoughts, and feel comfortable such that they will not be pressured by management or the client to present a professional image. This will likely need to be started by the Scrum Master. If this isn't part of the team culture, they will notice if the Scrum Master is taking the first risk in expressing a potentially contentious viewpoint.

Did the management make a decision you disagreed with? Why? Is the client being unreasonable? Was the last company meeting a huge waste of time? Share your perspective with the team first and they will likely begin providing honest feedback but more importantly, they can also begin to understand each other's perspectives too.

Identifying when direct involvement is needed can be difficult for the Scrum Master and others to determine effectively. When this is misunderstood or misapplied, this can set a negative precedence for Scrum Teams, causing even more work later to undo the poor practices. Examples include a Scrum Master that maintains the Backlog, writes every User Story for the team, provides constant technical input, and constantly asking for User Stories to be updated. Scrum Masters can end up in this situation and give in to pressure that they need to be directly delivering more so they can have a more concrete discussion with their manager. The Scrum Master will need to demonstrate confidence in their coaching ability and judgment of when to stay out of being directly involved. This may require coaching to managers as well if they continue to push the Scrum Master into unfavorable situations for the team. Not all Scrum Masters do this but they absolutely should. We should always be questioning what we're doing and why we're doing it. We should always be examining the impacts of our actions on the team. That's what it means to be a coach [9].

"An individual step in character training is to put responsibility on the individual." - Robert Baden-Powell

Coaching the delivery team is a core responsibility for the Scrum Master, however I'd include coaching managers that are involved with the team. Without guidance for the team lead or manager on the goal of growth for the team members, they can absorb too much responsibility, exhibit initiative and action rather than delegation or training, and can potentially be a dreaded micromanager. Understanding this and then fostering initiative, from the 11th Agile Principle, can reduce the stress all around the team. This is done with two actions at the same time. First, ask the lead to back off a bit on some of the responsibilities and perhaps ask them to delegate those responsibilities. On the other hand, work with the team members to focus on taking opportunities to step forward and have more ownership. Once team members have a taste of initiative, this will likely snowball until they reach their own limitation on what they're comfortable with doing.

Trust as the 5th Agile Principle is coached, not pushed. Direct involvement typically removes opportunities to establish trust in someone else. Trust is cultivated when the team relies on each other in the face of uncertainty. This largely benefits team members that are risk averse. Introducing very small changes incrementally is the best approach in sparking the risk tolerance of uncertainty. This requires building upon initial credibility and reminding the team that any changes are not expected to be a permanent solution, therefore take a risk and try something to see the results. They are always welcome to go back to how things were before if any new process fails to meet expectations.

Technical excellence from the 9th Agile Principle is collaboration and working towards getting the right questions asked. If situations are being resolved directly by a Scrum Master, or the IT manager for technical issues, then it's undermining the opportunity for professional development. It's fostered when team members are speaking up and challenging assumptions that may exist in designs and User Stories. Facilitating healthy conflict is the best way to ensure that the best approach is the one being taken. There is little direct action that can be taken to support this other than getting the right people to proactively and candidly talk to each other.

It can be great to sit with someone one on one to have a chat about their career and how they want to develop their experiences. Try to understand how they want to interact with others. Who do they want to be, an IT manager, a Product Owner, a specialized developer, an architect? What kind of engineer? Do they have Scrum experience, how was it? What kind of work environment do they want to have? This process is for them and the changes that will occur are not passive nor only completed by the Scrum Master. Having the team members share their vision with you, or possibly even prompting them to begin thinking about what that vision could be, will help successfully create a lasting, beneficial culture change. This is coaching over direct involvement as it also serves as an opportunity to encourage someone to pursue what they want, rather than having someone explicitly hand them an opportunity later.

An engineer I once worked with said that he felt stunted in career growth and wanted to experience the design phase of projects more. A broader exposure to the SDLC is always a good thing but he was particularly motivated for it. We broached the topic with the manager who was handling all of the designs. We figured out a way to include the engineers and eventually hand over the responsibilities when it made sense to do so. It would be a rotating schedule for the engineers to participate in the design process. First it would be observation, then contribution, then ownership while the manager would remain available for guidance. This added tremendous value to the User Stories downstream as the engineers themselves were able to ask more appropriate questions about expectations, as well as answering each other's questions. This helped remove the manager from being a bottleneck while also fulfilling the ambitions of the team. They both wanted this improvement to occur but sometimes it takes a little push to see it through.

Control your reactions well, especially if working with people new to Scrum. The way you react will set their expectations of how to work with the rules and guidance of Scrum. By keeping it relaxed, even if it's a critical issue that comes up, stress will be lower and everyone will be able to think more clearly. The odds are that the team is not working on embedded pacemaker software where lives are at stake, so it's just software that's relatively easy to fix. They will more readily understand the significance of accepted risk yet not treat it like the end of the world nor start to blame each other for fallout. By remaining level headed and understanding that Scrum rules sometimes need to be broken, then the team will understand that it's okay to think for themselves since the Scrum Master is approaching it very fairly too. The alternative is blindly applying rules and getting frustrated when it doesn't work out. You are seen as an authority on the process and a coach. The team will notice how you react to different situations and information coming your way.

Servant-leadership does not need to be limited to just Scrum Masters themselves. Anyone can demonstrate these qualities. It can be beneficial for the team, managers, and PO's to recognize the qualities when they're engaged by a Scrum Master and even demonstrate the style themselves. These qualities are identified as [10]:

- Listening
- Empathy
- Healing
- Awareness
- Persuasion
- Conceptualization
- Foresight
- Stewardship
- Commitment to the Growth of People
- Building Community

For coaching these qualities, the Scrum Master must be consistent in their application in a genuine way. Not everyone will be receptive as it often takes a natural inclination to serve to be interested in it. Some people will be interested, some will not and that's fine. As stated by the originator of servant-leader, Robert Greenleaf, to determine if you're having an impact through your efforts, "The best test is: do those served grow as persons: do they, while being served, become healthier, wiser, freer, more autonomous, more likely themselves to become servants?" [11]

Coaching to affect culture change is indicative of the careful discipline necessary as a Scrum Master. Upper management will likely have the assumption that you will step in and exert control over the team to affect immediate change. The team will probably expect this too, even if they've Scrum training and experience. It's vital to clarify with the team your approach as a Scrum Master because it's likely some of the team will not have worked with the role fulfilled before. There's a juxtaposition in the notion of how to affect culture change. You will need to stir a sense of dissent and open discourse on issues impacting the team to have any profound and permanent shift on the individual team member's way of working, rather than becoming a dictator. If you are only forcing your process on the team and they are merely compliant, as soon as you leave, they will go back to previous habits. This is why I propose Coaching over Direct Involvement. We want self-reliance, not compliance.

Story

Spring coolness fades into summer in the north eastern city. The trees and plants bloom into a fresh green. Flowers continue to emerge through the season. Tourists visit the city and snowbirds come back to roost after evading the winter storms. As the seasons change, so does the company Michael was recently hired to work for six months ago. Michael's time working with the single team has proven a positive experience so far.

Wearing summer attire allowed after Memorial Day, Michael stands at his desk as the team begins their Standup. He shares his screen on the wall-mounted TV. The team gathers around it and begins unprompted. They discuss their impediments and intent to resolve them. They finish in under 10 minutes with everyone feeling informed as to where they are in the Sprint. They ask if Simon has anything to add in Standup for the team while they're all together. Simon fills the team in on the details of the last major corporate communication.

FancyCat Technology has recently been acquired by a larger company in the industry, BiggerFish Corporation. As part of the merge, Michael will be taking over a new team with a new process. Before that had even started, however, the new corporate culture was immediately apparent. With each level of management, each manager wants to have their own monthly, or quarterly, meeting. In addition to all of these meetings, there are chapter meetings, communities of practice, general technical knowledge sharing, and other departments wanting to include everyone in every status update meeting.

This is overwhelming for the team.

After some speculation on how these merges might impact the immediate time, Simon asks, "All things considered, we're all still employed and working on our current code base. As for the meeting overload, we will likely just send one engineer at a time to represent the team, so they can pass on any significant information," he rolls his eyes, "if any exists there."

Erik raises his hand, "Hey Scrum Master, with all of these meetings and training occurring, I am getting burnt out on information. How do you expect any Scrum info to stick with the new team?" Michael thinks for a moment and replies.

Anthony Belau

Chapter 7
Conversation over Training

"Tell me and I forget, teach me and I may remember, involve me and I learn." - Benjamin Franklin

Conversation and communication can be sensitive soft skills employed based on the audience. These are rarely provided guidance for Scrum Masters. Conversation is certainly an art [11]. Although some people seem to be natural in navigating them, it is certainly possible to learn to become more proficient in communication for those less inclined or socially savvy. Considering that the Scrum Master has no specific authority, your ability to drive change and encourage teams is through garnered influence and credibility. While conversation is a strong way to create change and collaboration, typical training is also an option if used correctly, although it should be sparingly.

Training is typically formatted as a presentation of information. It is great for communicating a lot of information in a short amount of time. It also helps get everyone the same message, although comprehension may differ. Conversation will have tangents or be difficult to follow if everyone is not as knowledgeable about the material discussed. If certified as a Scrum Master, you can check if your certification accreditation source has any training materials available. If not, your company or other Scrum Masters may have presentations available. If they do not, you will have to start from scratch. Fortunately, we live in an era with incomprehensible amounts of free information readily available through blogs, online videos, free courses, and forums where many questions have already been asked and received various opinions so that any training presentations can be created from scratch.

When you're performing training, it is beneficial to have one of the first slides to be opening the floor to questions and letting everyone know that interjection is expected. It has been my experience that people are the most receptive to answers when they ask a question, especially if it's one that's unprompted. This can indicate that they've been thinking about it for quite some time if it's a Scrum question. It also sets a tone of openness about the topic and your expectations for communication. People should feel confident in their understanding of a subject. They should not rely on the presenter to offer all information nor make assumptions to fill any gaps.

Despite the value of training, when someone asks a question, it is then that they are most interested in the answer and information. This suggests when the most opportunities you will have to productively communicate on a relatively pedantic subject. Knowing this, most training I do occurs situationally or during impromptu discussions. Being focused and prepared every day in every conversation is imperative to being able to communicate effectively in these moments. If you are unable to answer a question on the spot, do not fail to follow up afterwards once you have had an opportunity to determine an answer.

Learn to lead with a compliment. We are working with individuals that have probably travelled a long, hard path of at least several years to get into the team. It's likely they did not receive a lot of compliments on their individual progress since they tend to be high performing individuals that have high expectations placed on them already. Some companies take the motivation of engineers for granted, not intentionally, however any relationship can become complacent this way. We should actively be aware of this and genuine in expression for appreciation and compliments.

- "Good question!"
- "Awesome work on getting that Story wrapped up last minute. Going forward, how can we avoid needing to do a huge effort at the end of a Sprint to finish it next time, if it happens again?"
- "I appreciate the initiative the team has been showing. I'll start to back off more in the meetings so you can have more ownership to direct it yourselves."
- "The open feedback during Retrospectives is great. I hope we can continue to have these discussions as things come up that impact you, even if it's something you think we can't fix.
- "The communication seems to be smooth between QA and Dev. Is there any chance we actually need more of it still or is what we have working?

Do not add the word "but" after a compliment. Many people do it and they mean well by it, however it can be interpreted like a backhanded compliment on the receiving end. It can feel disingenuous [12]. This is why the Scrum Master needs to take time to reflect and think about the context of situations, how things are going, and what they're going to say. It's better to say nothing than to say it wrong. Communication can seem a very simple topic, however you cannot often sincerely take words back once spoken.

A good problem to have is if your team merits many compliments for their hard work and responsiveness to critical issues, however too much of a good thing can always turn it sour. It can be perceived as disingenuous or belittling, although it is not the intent. When communicating, it's largely about the audience's expectations of the coach, not just what the coach wants to communicate. When the team does well by your expectations, that could simply be the norm for them and they are capable of achieving even greater results, should they need to.

It is best to start by asking simple, vague questions as the coach, no matter how obvious the answer may seem. Ask questions to the team even if you already know the answer. This puts the responsibility of knowledge and Scrum process ownership on the team rather than just the Scrum Master. Ask questions but don't drag it out if no one knows. Be careful to not dominate the conversation like it's an interrogation. Think of cop interrogation scenes in movies, CEO's asking their staff about status updates, or teachers asking questions of their students. Questions can create a power dynamic that's unintentional as a facilitator, depending on the presentation of the question.

- "It looks like Monday is a holiday but we've our Refinement on that day. How do we want to handle this?"
- "We've a large meeting overlapping our standup time. What did we want to do with it? Cancel or move the time? Why?"
- "It's great to bring this up in Retrospective, that the last minute Story was completed last minute. How

did we get there though? Should we expect to act differently if this were to occur again? Why or why not?"

To prompt impromptu conversation, just be your casual self. It's not beneficial to be a cold professional to the team. If they feel that you're distant and uninterested, they're unlikely to bring up any sensitive issues to you. Ask the team what they did over the weekend, offer some info about yourself, or about something that happened to you over the weekend. Crack jokes and laugh, it'll lighten the mood and encourage the team to do the same. Be very careful with sarcasm with the teams. It takes time to understand someone confidently enough to know sarcasm when you hear it.

Learn small talk and ask questions. As employees, we sometimes get influenced by the corporate concept that we're "human resources" or "human capital" but we are more dignified than that. Be open to sparking conversations and engaging people before meetings, afterwards, or even during. Ask about hobbies, travel, kids, plans for the weekend, experiences and opinions of Scrum, or just simply, how they're doing.

Be aware of how you are perceived. The Scrum Master may be perceived as a manager, which can impact openness. It's fair to explain that the next steps in growth will be largely determined by the team. This is not a dictatorship where you're setting all the rules. Some company cultures have an environment where the teams may attend a lot of meetings where they are not expected to contribute information or ask questions. This means that the introductory meeting is better if it should come across as conversational instead of instructional. Encourage the team from the beginning to ask questions and interrupt you. You are there to support the team, not to grade them, not to be an authority, and not to be a boss. People are most receptive to answers and suggestions when they are the ones to initiate the question. Four hours of training does not mean four hours of information has been confidently absorbed by the team.

There is no need to defend Scrum or Agile when it's criticized. In fact, I openly criticize the process for shortcomings and encourage conversation with team members about it as well. I remind them that these observations do not affect my performance or confidence as someone that's expected to be an "agile champion." Although we're responsible for the process, we didn't define its rules. We only work by trying to apply it to the team and negotiating its application so that they can become more effective in their work. This will open and encourage productive conversations closer to what the team truly feels.

It seems that people that gravitate to this role tend to be agreeable. The Scrum Master will need to be confident in saying "No" tactfully. This is necessary for working with team members, team leads, stakeholders, well-intentioned people within the business that may be used to free access to the IT team, and even your boss and your boss' boss. This can be important in relation to metrics and some information that management may ask about. You do not have to comply or provide it if it may ultimately be detrimental to the team.

For example, I had a manager ask for points completed per team member, which I refused to provide. The conversation didn't end there though. I instead tried to understand what he was trying to determine. In this example it's necessary to explain that when a team member is assisting others on the team, they will not personally complete the most points but they are enabling others on the team to complete more. Defending the team is critical to the preservation of trust and collaboration in this situation. This conversation came about as a result of well-intentioned expectations for a manager. More training on Scrum was not the answer, rather a conversation to get into the applied nature of Scrum in the self-organizing team.

The understanding of Scrum expectations and Agile values will vary widely and empirically for everyone. Expect to have your training clarify any discrepancies. Precise vocabulary is absolutely critical to explain consistently and provide relatable examples. A misrepresentation can have ramifications months down the line or cause arguments based on interpretations of what you meant to say. Be clear and succinct as less words in the training will add more value.

Learning to coach Scrum is like playing an instrument. A two-day certificate course may show you the basics of the instrument but you're only just beginning to learn how to play it effectively. Each Scrum Master will develop their own style and approach to situations, although definitions or holding to fundamentals should be consistently taught. Coaching Scrum is more successful by focusing on Conversation over Training. Be a Scrum Master, not a Scrum Pastor.

Story

BiggerFish Corporation has acquired a few companies recently. They're not hiring any Scrum Masters into the role. They've decided to internally promote someone. Bill, from another IT department, is currently working as Junior Quality Assurance. He's interested in trying the role but he has no experience at all with it. His manager has asked if Michael could get Bill up to speed. Michael is confident enough in his coaching abilities that he could mentor a new Scrum Master. He gives Bill a call after hearing the news.

"Excited about the new opportunity?"

"Absolutely! I'm not exactly certain about what I'm getting myself into though."

"No worries, if you have any questions or come across any situations, feel free to reach out to chat."

"Okay, great. I've done some reading on my own and I think I understand what Scrum and Agile is but I can't seem to get any info on how to actually do the job, like facilitation for the meetings."

"Yeah, it can be pretty situational."

"It seems like facilitation is more than just powerpoints and I'm not sure coaching is really what our middle managers do here. They just tell people what to do and ask for status updates but don't try to care about employees' career growth."

"Sounds about right."

"How are you currently running your Scrum Ceremonies? We've Sprint Planning tomorrow and I'm not sure what the team expects from me."

"Do you have the next hour free to go over some of it?

Chapter 8
Meeting Facilitation

A primary duty of the Scrum Master is meeting facilitation. This does not mean that the Scrum Master prepares all of the content for a meeting or controls who is invited. The responsibility is to see that a general purpose is provided and roughly followed, conversations remain reasonably productive while on task, and that it's timeboxed fairly, meaning it's ending when it's supposed to. You might be asking "Why can't a developer in the meeting just facilitate it themselves?" It seems obvious for someone to handle independently, however that's not necessarily the case in practice. Contributing and following along in a meeting discussion can be like reading a good book. The material can be engrossing such that you lose track of time or what the individual characters were initially supposed to be doing at the beginning of the chapter since something more significant came into the picture as the topic changes.

Sprint Planning is a meeting the Scrum Master owns and facilitates. There's often enormous flexibility in implementing it based on what I've seen in organizations. It marks the beginning of a Sprint and should be timed accordingly in the team's schedule to reflect that as closely as possible. It is ideally performed the morning of the Sprint start but some teams find it beneficial to do the evening before. Arguably it's a bad practice against the dogma but if it motivates the team and shows that their feedback is turned into action, the Scrum Master will have to make a choice that no one else can judge. Change the agenda based on the team's needs.

I've had Sprint Plannings range from 15 minutes to 2 hours, sometimes consistent, sometimes sporadic. Remember that the meeting is not for you. It is so that the team members understand what they're committing to and allow the PO to make any last minute adjustments with the team. The purpose is to walk away with a confident commitment, not rush and force the conversation into an arbitrary timebox.

The conversation pieces I often include are:
1. What were the metrics from the last Sprint?
2. Anything from the PO or updates to the Backlog needed?
3. Show the team their velocity.
4. Account for PTO and holidays to determine the capacity they could commit up to.
5. Have the PO walk through what they're hoping for in the next Sprint.
6. Team discusses and determines what they can commit to.
7. Have the PO rank and adjust anything the team cannot commit to for the next Sprint.
8. Discuss any risks or questions.
9. Any general questions?
10. Start the Sprint.

Standups are the daily meeting that's owned by the Scrum Master. It's imperative that the intent is understood by the team from the very beginning. This meeting's purpose is for the team to synchronize with each other, ask relevant questions, and comment on availability. Initially, some structure will need to be set by the Scrum Master but it can eventually be turned over to the team such that they will begin unprompted. The facilitator is expected to pay attention to what is being said and what is not being said. Sometimes understanding the content is necessary, sometimes it's not. What you're listening for is if people are communicating actual progress, if they're doing the same task without providing more insight into what's going on, and if impediments are actually being discussed.

The three key points:
1. What did you do yesterday?
2. What will you do today?
3. Are there are impediments?

Through facilitation, you can often ask these common questions:
1. Is there anything anyone can do to assist you moving forward?
2. Is there anything you expect to need sooner than later?
3. Of the User Stories we have remaining, are any at risk of not being completed? Why?

For Refinement sessions, which were previously referred to with the deprecated term Grooming sessions, this should be expected to be owned by the Product Owner. Often with an adoption of Scrum and Scrum Masters, Product Owners will be new as well. Guidance on facilitation and managing expectations, and keeping the conversations focused is critical for this meeting since this is where a majority of technical understanding occurs for each User Story. This is driven by questions from the team before they offer an estimate for the User Story. The team will eventually mature in how they ask their questions and the PO will learn what to expect from the team to come more prepared in future sessions.

It's essential to timebox the meeting. Depending on the team's immaturity, the Scrum Master should highlight when they're spending too much time on a Story and ask the PO to come back with more details for the next session. It has been my experience that when a team sees a Story for the first time, they will immediately have an estimate in their mind if the Acceptance Criteria is well written, which will sway very little based on conversation unless a particularly insightful question is asked. Getting a few questions answered first then estimating sets a good pace for getting through the Backlog. Five minutes per User Story is a good goal.

As a facilitator, ask these questions:

1. What do you need to know before you can estimate this Story?

2. What questions do you have that will affect your estimate?

3. Are there any scenarios this Story may miss?

4. What could be done to reduce the size of this Story?

5. What do we need for this Story before it can be started?

6. What do we need for this Story before it can be completed?

Prefinement is not a Scrum meeting. Most teams will perform the essence of this meeting. The name was coined by one of the developers at a company I was with when the PO needed to socialize some project ideas and User Stories to get them ready for Refinement. The socialization would prompt questions earlier in the process so the PO could be more prepared to have the Stories sized. We started with just the PO, lead, and one developer attending. Attendance of a developer would rotate through the team so the designing aspect of the process was more distributed and growing the cross-functional capabilities of the team. I had only facilitated a few of these early on as the lead provided most of the direction on what was needed. Although teams may be new to Scrum, they are not new to software development. This served as a great structure for them with a regular cadence that would keep the backlog healthy.

Sometimes work comes through with a technical description but that is not always the best way to communicate nor design something. The team will be closer to the code and more familiar with the application's expectations. They need to know what the general problem is that the business wants solved so that they may design the implementation.

Questions that are helpful in Prefinement meeting:

1. How does the priority of this compare to what we currently have?

2. Should we expect to work on this in the next Sprint?

3. What is the problem that is trying to be solved with this project?

4. Are there timelines and how flexible are they based on the solution needed?

5. Do we intend to use this functionality for anything else going forward?

6. Are we expecting to work with any other teams for this?

Sprint Reviews signify the end of a Sprint. This is the official Scrum opportunity for the PO to see what was completed. The agenda usually includes a functional demo of what was completed, a review of the metrics for the team, and a chance for feedback on any adjustments to be made to the User Stories. It is also a chance for stakeholders and consumers of the delivered product to see what was completed. During this meeting, regardless of the attendees, it's important to be very aware of how the team is represented and how feedback is received.

Sprint Review is often too late to receive practical feedback. Teams I've worked with find it best to keep the PO involved throughout the Sprint and actively demo their work as they're progressing within the Sprint. They also preemptively ask the PO questions on which direction to take as they become more familiar with designing code. The role of the Scrum Master is to establish this closer collaboration. Often a PO will be overloaded and not understand the priority of getting the team's questions addressed, however by getting the team to discuss the benefits during a Retrospective, this will hopefully establish improvements in this area. By allowing this kind of demoing to occur instead of forcing demoing to only occur during the Sprint Review, you're enabling the team to bring better value rather than shutting down collaboration for the sake of a prescribed Scrum process.

This meeting can sometimes be a sensitive meeting for the team. As they go through the stages of team development, they will sometimes do well, and sometimes not do so well. Teams are aware that their image can be important to those outside of their immediate circle in the company and can sometimes feel that they're compared to other development teams. If this feeling exists, do your best to soften it. As always, metrics are only relative to the respective team and cannot be used to compare teams. Metrics are also only the start of the conversation for what is occurring. Numbers themselves don't mean much other than a talking point to create transparency. They are a means to understanding. If an organization disagrees with this notion or has a culture that doesn't understand, it is expected that a Scrum Master will have to work to change this.

For example, when discussing velocity of teams A and B. Team A has a velocity of 20, Team B has a velocity of 60. This does not mean that Team B is doing three times the work of Team A. The 20 points that are completed in a Sprint for Team A is based on Story Points, which are an estimate of effort for a Story based on however the team gravitates to their own normalization. Some teams have a lot of 1's, 2's, and 3's, some teams have a lot of 3's, 5's, and 8's. If anything, the team that averages larger Stories is dealing with a lot more unknowns in their work while the team with smaller Stories is likely to be more consistent and accurate in their delivery of value. A simple solution is to exclude showing anyone outside of the team and PO what the team's numerical velocity actually is.

When it comes to metrics, I often ask, "What are we trying to determine with this information? What do we want to know? Why?" If the metric provides no transparency or discussion points, consider discontinuing it. A solid metric that can be called "Total Story Points Completed" is the percentage of the total Story Points accepted compared against the total Story Points committed. If Team A committed to 20 Story Points but during the Sprint, they broke commitment and brought in work resulting in 21 Story Points completed, this would be 105% Total Story Points completed.

To show the broken commitment, another metric called "User Story Completion" can be used. This is identified by the total number of independent User Stories committed to, then compared against how many of them were completed. For Team A, if they committed to four User Stories at 5 Story Points each for a total of 20 Story Points but then broke commitment and removed one User Story worth 5 Story Points from the Sprint and then pulled in two User Stories at 3 Story Points each, then of the four originally committed User Stories, only three were completed resulting in 75% User Story completion, although the team finished the Sprint with five User Stories completed and 105% Story Points completed. This helps demonstrate that plans changed but the capacity was still met for what the team thought they could do.

The Retrospective is the cornerstone of Scrum. This is where the feedback loop of incremental improvement occurs. Any other meeting can be pretty flexible but the Retrospective meeting needs to occur every Sprint. The Scrum Master's facilitation skills need to make it effective. If the Scrum Master is failing here, they are failing to implement Scrum regardless of everything else that may be going well. There's the exception that a team could be under an intense time crunch and need all available time to finalize their work. Of course this happens but if the team is under this kind of intense pressure, they will need to work to understand why it's happening and how they can mitigate this pressure going forward.

There are many templates available online but the standard one has been incredibly successful for me to work with.

The three basic points to address are:
1. What went well?
2. What didn't go well?
3. What can we do to improve?

The wording is important to consider as it will frame their perspective of the discussion. I had one team that was doing "What went well?" and "What can we improve?" This excludes the whole aspect of identification of problems that may seem unsolvable by the team. The same team had a process where the Scrum Master's facilitation was not through conversation, but rather through a 5-10 minute silent timebox where people would add notes, causing isolated thinking and no collaboration growth or real communication. I changed this immediately when I began working with them. I started receiving twice the feedback with constructive engagement mostly prompted amongst themselves, as well as ownership over solutions to implement.

My approach is that the Scrum Master acts as a facilitator and scribe during the meeting so only the Scrum Master is writing notes down while everyone else is free to focus on the conversation only. This forces the team members to verbalize their thoughts and concerns rather than anonymously offering complaints. They're also initially directing their attention to the Scrum Master rather than to someone that may have caused the issue so it feels more neutral for them to communicate their concerns, creating an environment that allows more initiative and risky opinions to be prompted. Often the team will bounce between the three topics until a major point is hit that will consume half of the meeting. As the scribe, the notes should be higher level on what concepts were discussed and asking for clarification on if what is written makes sense. It helps tremendously to ask clarifying questions so that the engineers will provide more details and context. Comments tend to focus on the symptom instead of working towards the root cause. I set aside a full hour, although sometimes it will end in 30 minutes if there's little else to discuss for the Sprint. For teams new to my facilitation and this meeting, I'll assure them that the results of the Retrospective are for the team's eyes only and that they're not shared with anyone else.

1. What about this went well specifically?

2. What was the circumstance that allowed this success?

3. How was it that the team was able to respond well to the last minute issue?

4. For this problem, what could we have asked earlier to potentially mitigate it?

5. Who can help us with this problem we're facing?

6. What is taking most of your time away from the Sprint commitment?

7. What is slowing you down the most in this work environment?

8. How was collaboration this Sprint?

9. Why did we miss this part of our commitment?

10. How can we ensure this doesn't happen again?

For example, a Retrospective topic went something like this:

"We were able to complete this User Story on time."

"Awesome, okay. Why was that?"

"Well the team lead helped us last minute to answer some questions we had to keep us moving."

"Interesting, so if he wasn't available, we might've missed that."

"Definitely."

"Right, so team lead, this is important because it shows how your availability and responsiveness is critical to our success, although it's not clearly defined in a metric nor is your capacity accounted for the way we consider the team's. How can we ensure this continues going forward?"

"I could have actually gotten involved sooner in the process but I was sitting in a few meetings that I wasn't really needed in."

"Okay, do we have the freedom to drop from company meetings that may not be important to us if it would help keep the team moving forward?"

"I'll check with the director. I think it should be okay but I'll try to be more responsive with the team if it helps that much regardless."

"Great! Next... How do we mean that we underestimated a Story in the Sprint? What happened? What questions did we miss in Refinement?"

In this example, we see how a simple thing that went well unfolded and what the expectations will be going forward to pursue more consistent success. Frequently, items called out by the team will be vague and generic. It can take some navigating to find more substantial information that the team needs to hear around an issue. I have tried other methods but I always come back to this approach as it's the best way for me to get the team engaged in the conversation. As a team matures, often a natural leader in the team will start driving some of the topics in Retrospectives after they see how unpacking a concern can bring about real value to themselves and the team. It's very rewarding to see the team solve their own issues after several Sprints under their belt.

The Scrum Master should apply tact and considerate timing when asking questions even with the best intentions. Interjecting in meetings, conversations, or focused periods of work will have a derailing impact and can negatively impact how you're received and lower your capacity for influence. Tact is necessary when following up on a complaint because it helps prevent conflict or drama by asking about a situation or the validity of a claim.

Not using tact when following up:
- "What are you doing all day?"
- "Why aren't you completing Stories as quickly as other engineers?"
- "What's taking so long to reply to the manager's emails?"

Using tact when following up:
- "Hey, just checking in with you. Are you having any impediments or need anything?"
- "How is everything going? Do you have any work that's not in the Sprint?"
- "Do you think we will have any issues meeting our commitment this Sprint? We've missed a few Stories from the last Sprint."

Following up without tact can make someone defensive if a negative point of view is assumed and the engineer is approached with it. When applying tact, understand that there is often a perceived concern that differs from what is actually happening. Always make sure that the engineer is fully enabled before assuming they're not doing work. Responses will more likely be a truthful answer and there's sometimes more to follow up with to help things move along. There are rare situations where the engineer simply isn't engaged and the perceptions are true, however as the Scrum Master, it's fair to always assume the best of individuals before believing a complaint to be true. By not following up on situations prudently, it can dissolve a sense of ownership over their work for the team member and contradict the viewpoint of a servant-leader.

During any meeting, it's better to leave more openings for others to speak up rather than pushing it along. I will count to five in my head after someone finishes speaking to see if anyone has initiative to jump in on their own. This aligns with the values of respect and courage, since without the opportunity to engage in the conversation, it's likely that there are fewer chances in other meetings not facilitated by a Scrum Master. As team members become used to speaking up, it's very likely that they may interrupt you and cut you off to ensure they're heard. I actively encourage this with anyone I meet when introducing myself or Scrum content. This helps so they know that if they think of something while I'm continuing the meeting, they're able to say what they need to say, while hopefully this carries over into other meetings where their need for understanding exists.

As a meeting facilitator, often questions are posed to the Scrum Master in the middle of a meeting. It can sometimes be difficult to think through a response on the spot since providing it since they're often a curveball or are about a situation that immediately revealed itself. Due to this, approach situations with an open mind, stay flexible, and express the fundamentals before going into context of the situation, it helps explain where the perspective is coming from. This also helps show that if a decision is based on fundamentals, it's one that they could be making on their own.

Reconsider how the Retrospective meeting is performed. I'm guilty of keeping the same pattern for a long period of time since it has been more beneficial than other methods. It can be helpful to add some sort of puzzles, challenge, or ice breaker at the beginning of the meeting to help break the continuity of the work day before reflecting on the last two weeks. Some teams have been around together but may not know the basic personal information of their team members. We spend over seven hours a day together as a team. On a given work day, team's probably spend more time together than they do their family. Generally, a simple ice breaker is all it takes for people to open up a bit.

Consider facilitating a social event with the team. This can build more team spirit. Set up a time to go out for lunch together. Set up time to bring in a board game. Learn a card game. Do an escape room. Push the team to attend happy hour, since IT can be notorious for never attending company events. Bring happy hour to IT itself and hang out with a movie on. Obviously, company and team culture determines the options available. This has only become more important in remote environments as companies close offices but maintain staff in the same region.

Chapter 9
Working with Scrum Masters

It is fair to assume that Scrum Masters will end up working without support of other Scrum Masters with the number of small companies that will attempt Scrum. The smaller companies will often pursue less experienced candidates for less pay to test the results of the process or to check a box in process adaptation. This is the area of opportunity for someone to gain experience in the role with a broader range of responsibilities that will be available. Scrum Masters are ambassadors of Agile and Scrum to organizations. There are also situations where Scrum Masters will be required to work in tandem with other Scrum Masters or in silos while still presenting a unified message to the teams. This will certainly take some managing for different perspectives and backgrounds as each Scrum Master's experience will be very unique with varying results through practices. Having access to individuals that can provide guidance into your specific situations is very helpful and the shared success becomes very rewarding as a Scrum Master team.

Expect different styles of communication. This role is highly dependent on personal experiences and successes. Some people will have software developer backgrounds, quality assurance backgrounds, business or sales backgrounds, military backgrounds, or simply no background because it's their first job. Every individual will have different ways of sharing their opinions, describing situations to the Scrum Master team, and perspective of coaching.

With the delicacy of establishing Scrum clout and rapport with teams, it is in the best interest of the Scrum Master team to be careful when criticizing other Scrum Masters and contradicting each other. Disagreements are better to discuss in private as an individual's ability to influence a team can be severely impacted by open contradiction. The team will consider the Scrum Master's perspective under the assumption that they know best in their domain of knowledge therefore it's safe to take a risk with them, but contradictions in that domain can make them more skeptical of their own Scrum Master. The most we should contradict is on an interpretation of how to apply Scrum and acknowledging it as such. Even if we feel confident in our perspective, it's best to err on the side of caution and assume the best intent, then follow up in private.

Start a Community of Practice at your organization if one does not exist already. The term sounds more formal than they end up being. A one hour meeting every other week has always been sufficient to keep Scrum Masters relatively connected, socialized, and to help coordinate training efforts. Having a group chat is also always helpful for keeping everyone synchronized on topics and question resolutions. As time continues, keeping this on a routine helps immensely for establishing the community. Often this will fall as a lower priority compared to other meetings, however it is valuable time spent when possible. I'll often use this meeting as a platform for mentoring a new Scrum Master and providing insight and questions on problems they're facing with certain scenarios they're seeing.

Problems pursued by the Scrum Master can sometimes be highly contextual, open to interpretation, and perspectives can influence the narrative. When a problem is presented, it's important to fully understand the situation more than just what the Scrum Master is describing. Ask questions and do not make assumptions. It's best to always guide to assume the best of a situation regardless of what the root cause is presented to be. We often do not have all of the answers and since we're expected to understand and analyze to create insights, it's a short step away from starting to make assumptions while speculating. When doing so, always assume the best case.

- Someone isn't responding to emails?
- They're not ignoring you, they're probably busy but check in on them.

- Someone isn't paying attention in Refinement?
- They're not bored or don't care, they're simply feeling like they need to work on something else instead. It happens but ask the engineer if he's able to focus.

- The team only completes 50% of their committed User Stories.
  - Maybe they're expected to be more flexible than Scrum would allow. It'll take a little digging to understand the nature of the expectations on the team based on the work coming their way.

If working with a new Scrum Master, it's a great opportunity to coach as well as collaborate. When an issue is explained, do not provide direct guidance on what to do next. Instead, ask them the high questions and see if they can navigate to the answer themselves. By helping them navigate the reasoning, they will be able to support their decisions later on their own, rather than saying "I was told to do this but I don't know why." If they're able to explain a decision and the reasoning against Scrum or Agile back to you, they're probably all set.

Coaching another Scrum Master has no guidelines on how to go about it. It's important to approach this subject with an open mind, no preconceptions of a situation, and to do your best based on the information at hand. Being a Scrum Master is difficult, a little empathy will go a long way in understanding some struggles and insecurities they may be facing. Being aware of Imposter Syndrome as a phenomena that does affect Scrum Masters as well since new ones are placed in a position of enormous responsibility, given some vague rules to follow, and are expected to achieve measurable results [13]. Any additional awareness of differences based on what the company needs for the role can help determine coaching needed so that everyone is as fully enabled to do their best.

Chapter 10
Closing

The question to ask is, "What are you doing that helps your teams succeed?" This role is not for the faint of heart but it certainly is very rewarding to know you had a hand in evolving team members, managers, and those around them to become lasting relationships. You will find that impediments raised by the team will incur quite a bit of weight for you to help get them resolved or escalated. You should feel empowered to pursue the larger issues up the organization until it can be resolved or at least rock the boat to bring awareness.

History often repeats itself and Scrum is not exempt from this idea. When Napoleon Bonaparte was campaigning and conquering much of Europe, one of the countries he fought was Prussia, modern day Germany. Napoleon is renowned for his ambition and innovative strategy [14]. One of the revolutionary ideas that gave him an edge over his opponents was how he rearranged his leadership structure for his officers [14]. During this period in history, militaries were organized with a command-and-control style leadership where the top officers on the field made all of the decisions and required a line of communication to each regiment to execute that order exactly with little room for the receiving regiment to apply their own judgment. One of the shortcomings of the style was that officers of these units were unable to make decisions on their own when confronted with a losing scenario or to capitalize on an advantage that may not be noticed by senior officers that could disappear given the lengthy turn-around time on communicating orders.

Napoleon turned the traditional dogma on its head when he empowered his officer corps to make decisions on their own, albeit within certain parameters and a vision of the overall goal. The parameters might've been something like they could not be more than a day's march away or seem ready to attack at a moment's notice. This enabled his officers to act independently and take strategic advantages based on the current information that they had and to capitalize on immediately. If they needed to relay information, they could lose the opportunities. Written communication which isn't always reliable either. One of the most vaunted cavalry charges in the 19th century, the Charge of the Light Brigade, was a direct result of horribly failed interpretation of a written order causing devastation to the cavalry unit that attempted to attack a well defended position [15]. This new approach in leadership and trust at the lower tiers caused massive wins for Napoleon [14].

Prussia had to completely reform its military to keep up with the agility of Napoleon's forces. This involved mirroring his command structure because theirs was demonstrably insufficient and too slow to react against the independent judgment of the French officers. Companies are struggling with the same concepts after 200 years in the IT field, where only the agile will outlast their competition in the modern marketplace.

Being a successful Scrum Master, like any other position, takes commitment and substantial effort. Should you choose to pursue this role, I have unwavering faith in your ability to resolve the challenges that will be set before you as I have seen countless others persevere in the face of extreme uncertainty by the individual. The opportunities within this position are incredibly wide and rewarding. Some will cater to those that would like to help learn to code or quality assurance. Some will expect more coaching or innovation. The industry needs more awareness about this position as a great opportunity. I hope that based on these applied experiences, you will decide to step into the world of Information Technology to see where that path may take you.

Chapter 11
Acknowledgements

Thank you for being interested in being a more effective Scrum Master. I hope that my applied experiences as a servant-leader and coach prove to be useful wisdom on how to navigate challenges that you will face, should you decide to perform this role temporarily or as a career. Coaching software development teams can be turbulent provided the vagueness of the role's description and breadth of assumed potential responsibilities, which varies wildly between companies. However, given enough introspection, collaboration, and patience, you will certainly come to the right approach to any problem.

Thank you Tony Galluscio for hiring me into my first IT job as a Software Tester at Veratics for the team which I later became the permanent Scrum Master with. Thank you Jeremy Noel for not laughing at me when I referred to a basic workflow chart as an entity relational diagram when our team was still forming and for not tiring of my unrelenting jokes. Thank you Michael Clancy for providing your technical expertise and professional wisdom to Team Echo with Jeremy and myself.

Thank you Laura Spigone for bringing me into Worldstrides as the first hired Scrum Master into the growing company. It was an empowering experience to have that depth of responsibility and support to innovate and trailblaze with the Explorica team. Thank you Patrick Coraccio and Michael Casallas for welcoming me onto the applications sole team and for trusting my guidance as we successfully tried different Scrum practices with the developers, one Sprint at a time, and adapting based on Retrospective feedback.

Thank you Kathy Shediac for bringing me into FIS and into the wide world of Scaled Agile Framework, SAFe. I thought I knew the totality of responsibility available for the Scrum Master but SAFe is much larger than I could have anticipated. Thank you Hillary Baird and Joanna McGarry for your open collaboration as we grow into the RTE roles for our trains as they evolve to become all they can be for FIS.

Thank you to the developers, quality assurance, and Product Owners who worked with me and committed to our practices through the business migrations, agile transformations, and trusting me to provide guidance through all of the stress. Thank you management, especially the ones I didn't get along with, for your patience and risk tolerance. Thank you to all military personnel and veterans and I hope you may consider a place in IT to find a continued purpose and meaning.

Chapter 12
Sources

1. https://www.historic-uk.com/CultureUK/History-of-Rugby-Football/

2. https://www.scrum.org/resources/blog/scrum-master-coach-0

3. https://www.investopedia.com/terms/p/pdca-cycle.asp

4. https://www.prweb.com/releases/neuroscience_based_fresh_tri_proves_iterative_mindset_drives_habit_formation_and_weight_loss/prweb16849730.htm

5. https://hbr.org/2018/05/the-surprising-power-of-questions

6. https://www.ccl.org/articles/leading-effectively-articles/coaching-others-use-active-listening-skills/

7. https://greatergood.berkeley.edu/article/item/six_habits_of_highly_empathic_people1

8. https://hr.mit.edu/learning-topics/teams/articles/stages-development

9. https://coachingfederation.org/

10. https://yscouts.com/10-servant-leadership-characteristics/

11. https://www.entrepreneur.com/article/326280

12. https://www.greenleaf.org/what-is-servant-leadership/

13. https://medium.com/@mcpflugie/using-the-word-but-a-brief-guide-to-sincerity-clarity-ab311a2af6a6

14. https://time.com/5312483/how-to-deal-with-impostor-syndrome/

15. https://core.ac.uk/download/pdf/36698436.pdf

16. https://en.wikipedia.org/wiki/Charge_of_the_Light_Brigade

Anthony Belau

About the author

Anthony Belau has five diverse years as a Scrum Master as the time of writing. His experiences range from serving a small, very high-performing team to coaching several application development teams as the first Scrum Master hired for Worldstrides. Before discovering Scrum Mastery, he graduated from the University of Central Florida with a B.A. in Interdisciplinary Studies and joined the Florida Army National Guard where he deployed to Qatar immediately after graduating. After coming home, he opted to go to Afghanistan as a civilian security contractor. These experiences provided a distinct perspective on selfless leadership, communication, and motivation. After coming home a second time, he began to learn Java for free through Youtube with Derek Banas' series. He was then hired at a very small company as an entry level Quality Assurance Software Tester, which segued into his first Scrum Master position after six months.

Anthony Belau